AF385726

*Current Topics
in Neuroendocrinology*
Volume 2

Adrenal Actions on Brain

Editors
D. Ganten and D. Pfaff

Contributors
T. Baker U. Beckford B. Bohus E. R. De Kloet
A. J. Doupe B. Gillham B. D. Greenstein
M. C. Holmes M. T. Jones B. S. McEwen
P. H. Patterson W. F. Riker, Jr. A. Sastre
H. D. Veldhuis

With 25 Figures

Springer-Verlag
Berlin Heidelberg New York 1982

Editors

Dr. DETLEV GANTEN, M.D., Ph.D.
Pharmakologisches Institut
Universität Heidelberg
Im Neuenheimer Feld 366
D-6900 Heidelberg/FRG

Dr. DONALD PFAFF, Ph.D.
Rockefeller University
York Avenue, and 66th Street
New York, NY 10021/USA

The picture on the cover has been taken from Nieuwenhuys R., Voogd J., van Huijzen Chr.:
The Human Central Nervous System. 2nd Edition. Springer-Verlag Berlin Heidelberg
New York 1981

ISBN-13: 978-3-642-68338-1   e-ISBN-13: 978-3-642-68336-7
DOI: 10.1007/978-3-642-68336-7

Library of Congress Cataloging in Publication Data. Main entry under title: Adrenal action on brain.
(Current topics in neuroendocrinology; 2) Bibliography: p. Includes index. 1. adrenocortical
hormones – Physiological effect. 2. Brain. 3. Glucocorticoids – Physiological effect. I. Ganten, D.
(Detlev), 1941. II. Pfaff, Donald W., 1939. III. Series [DNLM: 1. Adrenal cortex hormones –
Physiology. 2. Brain – Drug effects. W1 CU82Qv.2/WK 755 A242]   QP572.A3A37   599.01′88
82-754   AACR2

2121/3130-543210

Contents

Glucocorticoids and Hippocampus: Receptors in Search of a Function

BRUCE S. McEWEN[1]

Contents

Abbreviations

ACTH: adrenocorticotrophin; ADX: adrenalectomized; CNS: central nervous system;
GABA: gamma amino butyric acid; GPDH: glycerol-3-phosphatedehydrogenase;
^{3}H: tritium; VIP: vasoactive intestinal polypeptide

1   Introduction

In 1968 we reported that the hippocampus of adrenalectomized rats takes up and
retains ^{3}H-corticosterone (McEwen et al. 1968). This and subsequent studies, which
were intended to identify and localize sites of adrenal steroid feedback in the central
nervous system in terms of putative receptor mechanisms, were stimuli for a variety of
investigations into the role of the hippocampus in adrenal steroid feedback on ACTH
secretion and on behavior. In this respect, the identification of putative receptor sites
was the impetus for the search for the function of glucocorticoids in hippocampus and

---

1   The Rockefeller University, 1230 York Avenue, New York, NY 10021, USA

in the rest of the brain. That search now seems to be coming to fruition. This article is a summary of the highlights of the search and the new information and perspectives which are coming out of it.

2 Glucocorticoid Receptors in Hippocampus and Other Brain Regions

2.1 Retention of ^{3}H-Corticosterone by Hippocampus of Rat and Rhesus Monkey

Corticosterone is the principal adrenal steroid in the rat (Bush 1953). When adrenalectomized (ADX) rats are injected intraperitoneally with ^{3}H-corticosterone, there is a greater uptake and retention of radioactive steroid by hippocampus relative to serum, pituitary gland, and brain areas such as hypothalamus and cortex (Fig. 1). This pattern has been reported and confirmed a number of times for the rat (McEwen et al. 1968, 1969; Stevens et al. 1971; Knizely 1972). It has also been extended to the rhesus

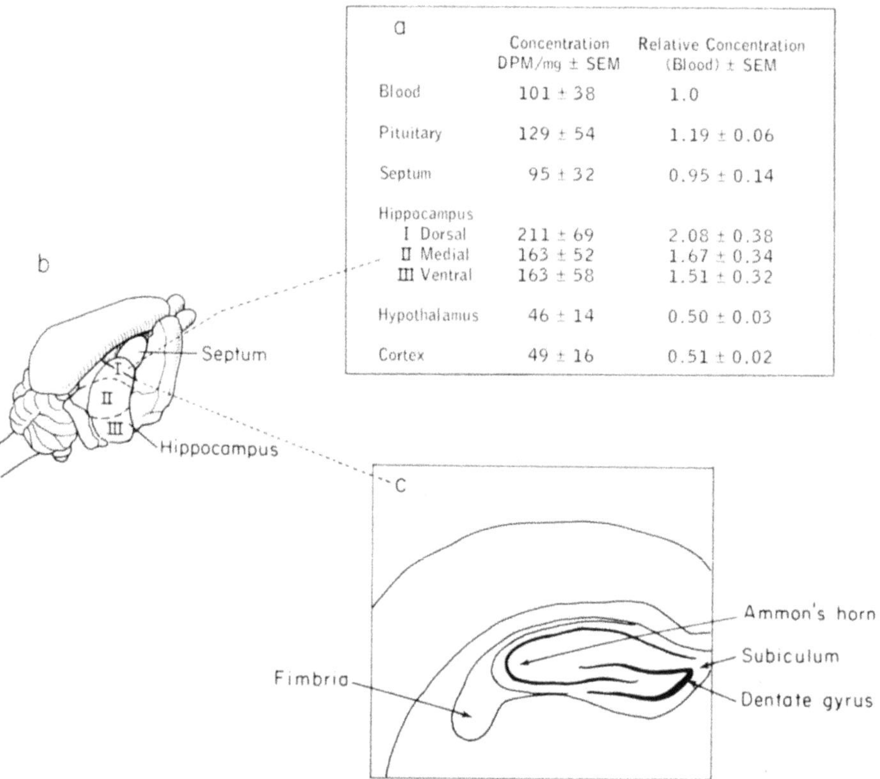

Fig. 1a—c. Hippocampal structure in relation to ^{3}H-corticosterone uptake. **a** ^{3}H-corticosterone uptake in dorsal, medial, and ventral hippocampus with other brain regions (see McEwen et al. 1969); **b** Position of hippocampus in rat brain, showing dorsal (I), medial (II), and ventral (III) portions. **c** Cross section of dorsal hippocampus showing cellular layers (see Fig. 2). From McEwen et al. (1972)

monkey (Gerlach et al. 1976), indicating that there may be a common plan of glucocorticoid-sensitive cells among mammals.

Autoradiographic studies after in vivo labeling of ADX rats with ^{3}H-corticosterone revealed that neurons are labeled (Gerlach and McEwen 1972; McEwen et al. 1975; Warembourg 1975a). Autoradiography of rat brains labeled in vivo with ^{3}H-cortisol also demonstrated hippocampal neuronal labeling (Stumpf 1971). Neuronal labeling is especially heavy in the pyramidal neurons of CA1 and CA2 of Ammon's horn and in the granule neurons of the dentate gyrus (Fig. 2). Precommisural hippocampus and indusium griseum are also labeled. Neurons of the anterior and lateral (but not medial) septum are labeled by ^{3}H-corticosterone, as are neurons in the basomedial and cortical regions of the amygdala. There are labeled neurons scattered throughout the cerebral cortex with a higher density being found in piriform and entorhinal cortex than in neocortex.

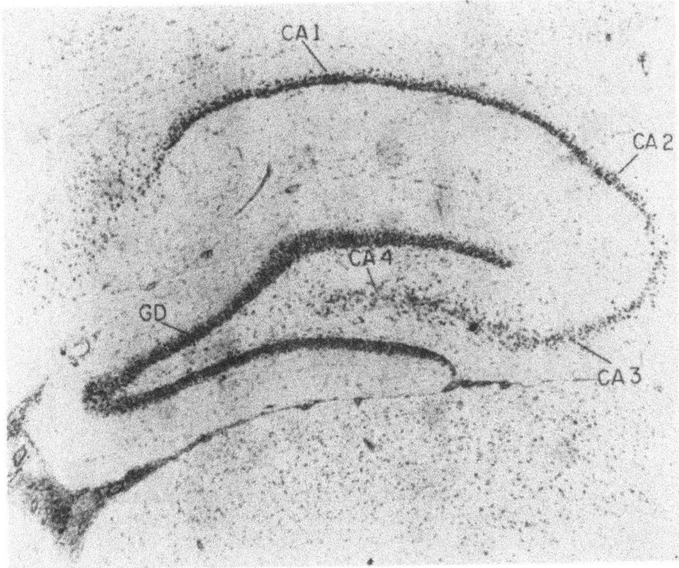

Fig. 2. Unstained autoradiograph of the dorsal hippocampus of an adrenalectomized male rat showing uptake of systemically injected (^{3}H) corticosterone as black silver grains. All contrast is due entirely to silver grains, which delineate the hippocampus anatomically. A sagittal frozen section was exposed for 608 days. The figure shows the longitudinal fields *CA1* to *CA4* of the pyramidal neuron layer in Ammon's horn, and the granule neuron layer in the dentate gyrus *(GD)*. Autoradiograph by John Gerlach. See Gerlach and McEwen for method (1972). From McEwen et al. (1975)

2.2 Cell Nuclear Receptors of Corticosterone in Hippocampus

Retention of ^{3}H-corticosterone by hippocampus is explained by the presence in this structure of soluble (Grosser et al. 1971, 1973; McEwen et al. 1972) and cell nuclear (McEwen et al. 1970; McEwen and Plapinger 1970) macromolecules with a high affinity for corticosterone. Cell nuclear retention of ^{3}H-steroids in vivo is an especially sensitive and revealing index of the regional distribution and steroid specificity of the glucocorticoid receptor system in the CNS (see below and Fig. 6). Furthermore, the measurement of cell nuclear corticosterone by radioimmunoassay made it possible to show that hippocampus is the preferential uptake site for the endogenous corticosterone secretion (McEwen et al. 1980) as well as for exogenous ^{3}H-corticosterone (Fig. 3).

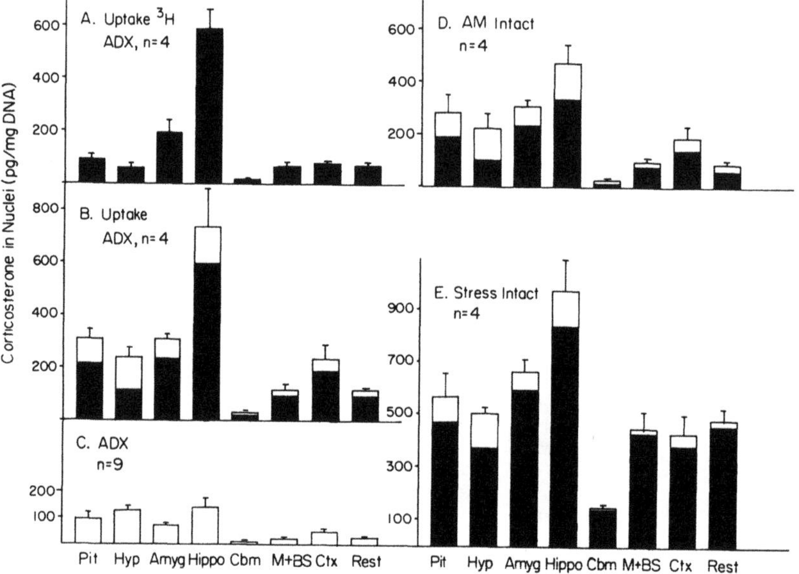

Fig. 3A—E. Cell nucelar corticosterone levels were measured in brain regions and pituitary. *n*, number of experiments on pooled tissue for five to six rats (**C—E**) and for three rats (**A** and **B**). *Pit,* pituitary; *Hyp,* hypothalamus; *Amyg,* amygdala; *Hippo,* hippocampus; *Cbm,* cerebellum; *M + BS,* midbrain plus brain stem; *Ctx,* cerebral cortex; *Rest,* remaining brain tissue. *Open bars,* total radioimmunoassayable material; *black bars,* corticosterone level with ADX values (**C**) subtracted. **A** Uptake of ^{3}H-corticosterone in ADX rats. **B** Uptake of unlabeled corticosterone in ADX rats. Note that pattern resembles that in **A. C** ADX rats. Identity of material accounting for values is unknown; it is probably not corticosterone. **D** Intact rats killed at 08:30 hours. Note that pattern resembles that in **A** and **B. E** Intact rats killed 15 min after a 1-min ether stress. Note that at high levels of corticosterone, nuclear retention occurs brainwide, possibly in glial cells (see text). From McEwen et al. (1980)

Cell nuclear retention of ^{3}H-corticosterone by hippocampal pyramidal neurons is evident from autoradiography (Fig. 4). Cell nuclear ^{3}H-corticosterone can be extracted from isolated brain cell nuclei with 0.4M KCl, and a substantial portion of this radioactivity remains bound to macromolecules, the putative receptors, during passage through a Sephadex G100 column (Fig. 5; McEwen and Plapinger 1970).

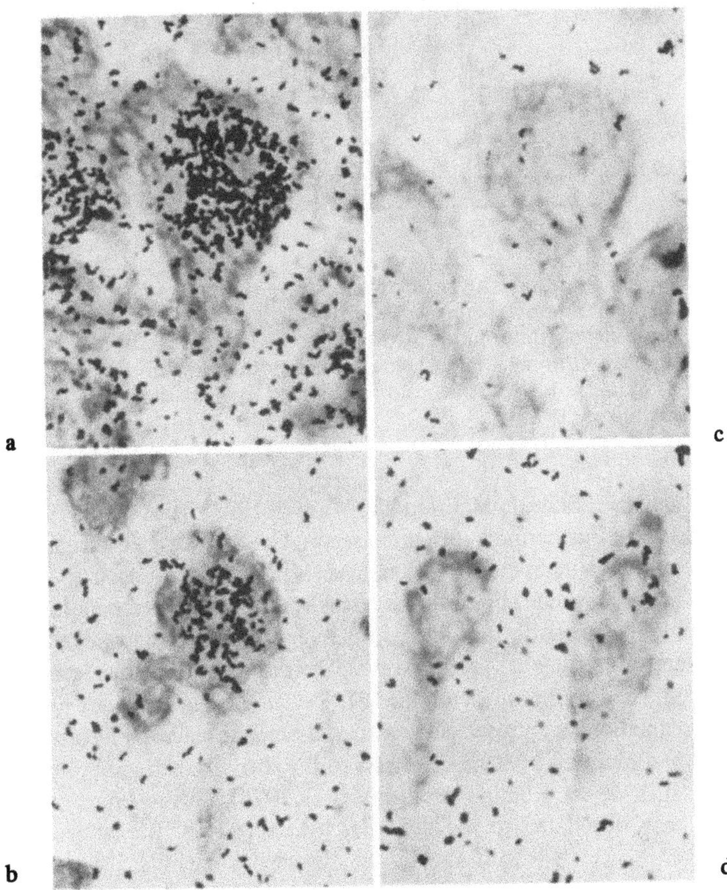

Fig. 4a—d. Autoradiographs of ^{3}H-corticosterone uptake by hippocampal neurons of adrenalectomized male rats. Method as described by Gerlach and McEwen (1972). a and b Control uptake of ^{3}H-corticosterone. c and d Uptake of ^{3}H-corticosterone in presence of 3 mg unlabeled corticosterone injected to compete for binding sites. (Photographs by John Gerlach). From McEwen and Pfaff (1973)

The relationship between hippocampal cell nuclear retention of ^{3}H-corticosterone and the binding of ^{3}H-corticosterone to cytosol macromolecules has been studied in some detail because of the possibility that cytosol binding may be an obligatory step prior to nuclear retention. The time course of in vivo labeling by ^{3}H-corticosterone reveals that binding to cytosol receptors precedes the elevation in cell nuclear radio-

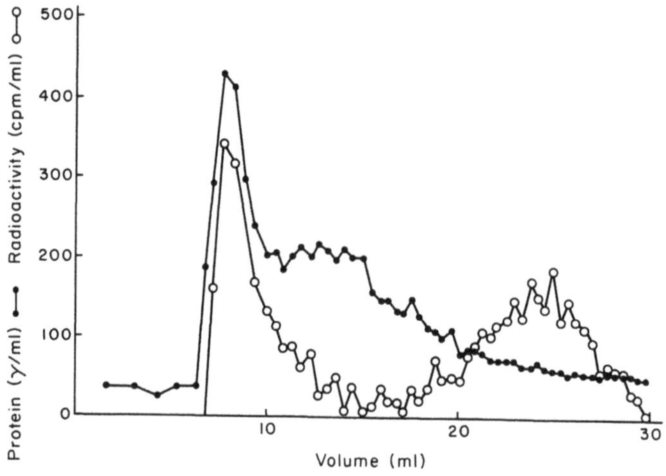

Fig. 5. Elution pattern of protein (●———●) and cpm (○———○) in a 0.4 M NaCl extract of rat brain nuclei labeled with ^{3}H-corticosterone-1,2 from a Sephadex G-200 column. Protein was determined directly on 0.1 ml of eluate and radioactivity was extracted directly from 0.1 ml of eluate with toluene scintillation fluid. The high baseline of protein is due to the color produced by the standard buffer. From McEwen and Plapinger (1970)

activity (Table 1; McEwen and Wallach 1973; Rhees et al. 1975a; Turner and McEwen 1980). Furthermore, cell nuclear uptake of ^{3}H-corticosterone in slices of hippocampus in vitro is temperature dependent (McEwen and Wallach 1973), much like the temperature-dependent cytosol-to-nuclear transfer of ^{3}H-estradiol in the uterus (Jensen et al. 1968). Moreover, labeling of nuclei by ^{3}H-corticosterone in slices is blocked by progesterone (Table 2), a steroid which binds to cytosol macromolecules but which is not retained by the nucleus (Fig. 6). Yet at the time of peak nuclear loading, there is no depletion of cytosol glucocorticoid-binding macromolecules in hippocampus (Turner and McEwen 1980), in contrast to the story for estradiol retention in uterus (Jensen et al. 1968) and brain (MacLusky et al. 1976). This unexpected finding for hippocampus may be related to the fact that the capacity of the cytosol macromolecules to bind

Table 1. Time course of cytosol and nuclear binding of ^{3}H-corticosterone in hippocampus of ADX rats. Data from Turner and McEwen (1980)

Time after injection [a]	Binding [b] (f/mols/hippocampus) in		
	Cytosol	Nuclei	Sum
15′	347	104	451
30′	326	180	506
60′	194	241	435
120′	92	402	494

[a] Dose 106 nmols/kg in tail vein

[b] Tissue pooled from three animals at each time point

Table 2. Competition by steroids for cell nuclear uptake of ^{3}H-corticosterone in hippocampal slices. [a] (McEwen and Wallach 1973)

Steroid	Concentration (M)	n	N
Control		4	25.9 ± 3.3
Corticosterone	2×10^{-7}	4	5.4 ± 0.4
Hydrocortisone	2×10^{-7}	4	9.5 ± 1.4
Progesterone	2×10^{-7}	5	21.0 ± 1.3
Progesterone	8.5×10^{-6}	4	5.7 ± 0.4

[a] N, nuclear concentration in disint./min/µg protein. ^{3}H-Corticosterone 2×10^{-8} M, incubation 30 min at 25°C

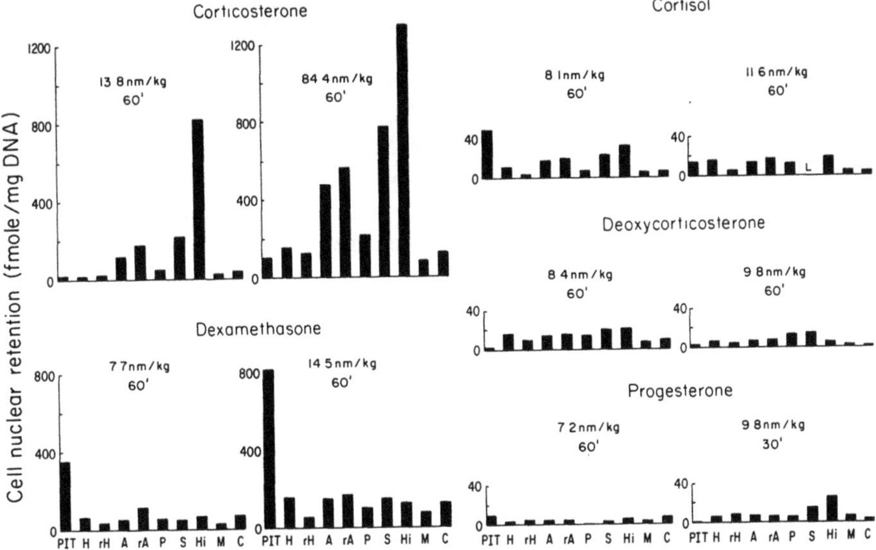

Fig. 6. Cell nuclear retention of ^{3}H-steroids by various brain regions after their infusion via tail vein in ADX-OVX rats. Time between injection and death and dose of ^{3}H-steroid (nmols/kg body wt.) are indicated for each experiment. Tissue from three to four identically treated rats was pooled for cell nuclear isolation. *L*, lost sample; *PIT*, pituitary; *H*, basomedial hypothalamus; *rH*, rest of hypothalamus; *A*, corticomedial amygdala; *rA*, rest of amygdala; *P*, medial preoptic area; *S*, septum; *Hi*, hippocampus; *M*, midbrain; *C*, cerebral cortex. From McEwen et al. (1976)

^{3}H-corticosterone in vitro exceeds that of the hippocampal nuclei by a factor of 2 (Turner and McEwen 1980). A large portion of the cytosol macromolecules are "cryptic" in vivo in that they are not occupied by doses of corticosterone which extensively occupy the cell nuclear sites; this finding and the above-mentioned lack of depletion with nuclear loading has led to the speculation that there may be, in the intact cell, a pool of cryptic receptors acting as a source of functional receptors, so as to maintain a constant level in the cell (Turner and McEwen 1980).

2.3 The Uptake and Retention of ^{3}H-Dexamethasone and Other Steroids

Just as the hippocampal uptake of ^{3}H-corticosterone was unexpected, so also was the subsequent finding that ^{3}H-dexamethansone, a synthetic glucocorticoid, was not retained selectively by the hippocampus but concentrated rather by the pituitary (De Kloet et al. 1975). Autoradiographic studies with ^{3}H-dexamethasone supported this (Rees et al. 1975; Warembourg 1975b; Rhees et al. 1975b). One study noted that after systemic administration, highest levels of ^{3}H-dexamethasone radioactivity occur near the ventricles, as if this steroid enters the brain, at least in part, by diffusion from the ventricles (Rees et al. 1975). Other ^{3}H-glucocorticoids and ^{3}H-progesterone are not markedly retained in vivo by brain region or pituitary in adrenalectomized rats (see Fig. 6).

The extensive uptake of ^{3}H-dexamethasone, compared to ^{3}H-corticosterone, by anterior pituitary has been linked to its inability to bind to the serum protein transcortin (De Kloet et al. 1975; De Kloet and McEwen 1976). Transcortin-like macromolecules are found in pituitary tissue, even after extensive perfusion to remove blood (De Kloet et al. 1977; Koch et al. 1977), and their presence may act as a barrier for the natural glucocorticoid, corticosterone, to bind to the intracellular receptors. Whatever the explanation, glucocorticoid receptor sites in pituitary are less extensively occupied by low levels of corticosterone than those in hippocampus (Rotsztejn et al. 1975) and are available for labeling by ^{3}H-dexamethasone in vivo unless very high levels of corticosterone are achieved (Koch et al. 1975).

2.4 The Identification of Glial Cell Glucocorticoid Receptors

Labeling studies with ^{3}H-dexamethasone reveal a rather uniform distribution of radioactivity across brain regions (Fig. 6), and autoradiographic studies with ^{3}H-dexamethasone point to a labeling of neuropil, although little, if any, localization to cell nuclei typical of glial cells was detected (Rees et al. 1973, 1975b; Warembourg 1975b). Yet biochemical evidence has for many years pointed toward glial cells as glucocorticoid targets (see below). The existence of glial cell glucocorticoid receptors has only recently been clarified and this information enables us now to better interpret the uptake patterns such as are presented in Fig. 6 for ^{3}H-dexamethasone and ^{3}H-corticosterone.

Glial cells have been linked to glucocorticoids since the report by DeVellis and Inglish (1968) that glycerol-3-phosphate dehydrogenase (GPDH) is regulated throughout the rat brain by glucocorticoids. A subsequent demonstration that C6 glial tumor cells of glial origin in culture show glucocorticoid regulation of GPDH (DeVellis et al. 1971; DeVellis and Inglish 1973) reinforced the suspicion that GPDH in the brain is a property of glial cells. This is now strongly supported by the recent immunocytochemical localization of GPDH to oligodendroglial cells (Leveille et al. 1980). Furthermore, glucocorticoid-induced hypertrophy of astrocytes has been reported (Scheff et al. 1980), as well as inhibitory influences of glucocorticoids on myelination (Friederich and Bohn 1980).

C6 cells contain glucocorticoid receptors (DeVellis et al. 1974), but until recently it was impossible to clearly identify such receptors in neural tissue. Recent studies on the optic nerve of the rat after eye removal have demonstrated cytosol and nuclear glucocorticoid receptors and GPDH induction by glucocorticoids in the surviving tissue (Meyer et al. 1982a,b). An interesting feature of in vivo retention of ^{3}H-glucocorticoids by optic nerve cell nuclei is that labeling by ^{3}H-dexamethasone is greater than that by ^{3}H-corticosterone. Thus it appears likely that the relatively uniform uptake of ^{3}H-dexamethasone across brain regions may be, to a large extent, a reflection of such glial cell uptake, whereas the regional variation in uptake of ^{3}H-corticosterone is a reflection of the presence of neurons which selectively retain this steroid. It should be noted that some limited neuronal labeling by ^{3}H-dexamethasone is reported for hypothalamus (Rees et al. 1975; Warembourg 1975b) and that corticosterone is undoubtedly bound by glucocorticoid receptors located in glial cells (see next section and Table 4).

Attempts to identify subtypes of glucocorticoid receptors in neural tissue which might represent glial and neuronal receptors have so far proved inconclusive. Many of the physical properties of glucocorticoid receptors from optic nerve, pituitary, and hippocampus (Meyer et al. 1982a,b) and from hippocampus and liver (Wrange 1979) are very similar, if not identical. The demonstration of two forms of neural glucocorticoid receptor distinguishable by isoelectric focusing (MacLusky et al. 1977) appear to be explained as the result of partial proteolytic digestion of a single receptor form (Wrange 1979).

The most likely explanation for the weak autoradiographic localization of ^{3}H-glucocorticoids in glial cells is that the concentration of receptors translocated to the glial cell nuclei is too low for autoradiographic visualization. The logic of this explanation does, however, leave open the possibility that many neurons not presently recognized to have glucocorticoid receptors, based on autoradiography, may in fact have some receptors. It would therefore be premature to exclude any brain region as a glucocorticoid target on the basis of negative evidence from autoradiography. Indeed, brain regions not heavily labeled by ^{3}H-corticosterone or ^{3}H-dexamethasone in vivo do contain glucocorticoid receptors when assayed in vitro by biochemical techniques (McEwen and Wallach 1973; Olpe and McEwen 1976). In such regions (e.g., cerebellum) the cryptic receptors (see above and Turner and McEwen 1980) are predominant. One explanation for the cryptic receptors might be a nonuniformity of access of circulating glucocorticoids to all neural cells. This might be due to a nonuniform diffusion of steroid from blood vessels, or even to a nonuniformity of blood flow. It might also reflect a route of access to brain tissue from the ventricles, as has been observed for ^{3}H-dexamethasone (Rees et al. 1975).

3 Glucocorticoid Effects on the Chemistry, Morphology, and Electrophysiology of the Hippocampus

How does the function of the hippocampus change as a result of the presence or absence of corticosterone? Answers to this question have been slow to come, but recent results show great promise. We shall first consider effects of chronic glucocorticoid treatment and then acute effects.

Chronic glucocorticoid replacement was greatly facilitated by the development of a simple method of replacement by means of subcutaneously implanted solid beads of steroid diluted, if necessary, with cholesterol (Meyer et al. 1979a). The initial studies with this method provided us with a list of some 16 enzymes of hippocampus from adrenalectomized rats not altered by corticosterone replacement for 1 week (Table 3; Meyer et al. 1979b). Many of these enzymes are involved with energy metabolism and the rest with biosynthesis or breakdown of neurotransmitters. The only enzyme that did change in this study was GPDH; this at least indicated the efficacy of the replacement therapy (Table 3). A subsequent study of GPDH (Meyer et al. 1979a) compared the efficacy of corticosterone and dexamethasone: 100% corticosterone pellets and 25% dexamethasone/75% cholesterol pellets produced or less equal elevation of GPDH activity in hippocampus (Table 4). Measurement of pituitary ACTH levels after such replacement (Gibson et al. 1979) indicated that 25% dexamethasone pellets were, if anything, more effective than 100% corticosterone in suppressing them (Table 4).

Table 3. Enzyme activities (μmol product/g protein/h) in crude hippocampal homogenates from adrenalectomized male rats with or without corticosterone replacement. Values are expressed as the mean ± SEM. Number of subjects is shown in parentheses. (Meyer et al. 1979)

Enzyme	Group	
	ADX	ADX + CORT
MDH	153 ± 5 (11)	153 ± 7 (8)
LDH	8329 ± 563 (10)	8964 ± 649 (8)
ICDH	273 ± 14 (11)	303 ± 12 (8)
SDH	124 ± 4 (11)	122 ± 4 (8)
G6PDH	98 ± 4 (10)	101 ± 8 (8)
6PGDH	136 ± 6 (11)	142 ± 8 (8)
GDH	1200 ± 130 (7)	1130 ± 100 (7)
GPDH	183 ± 17 (6)	481 ± 49 (6) [b]
HK	2698 ± 134 (11)	2823 ± 172 (8)
PK	2272 ± 77 (10)	2528 ± 135 (7)
GLN-ASE	695 ± 65 (6)	735 ± 60 (6)
GABA-T	580 ± 25 (6)	550 ± 45 (6)
ASP-T	11730 ± 400 (7)	12360 ± 980 (7)
GAD	202 ± 6 (6)	212 ± 6 (6)
MAO	45 ± 2 (6)	45 ± 2 (6)
CAT	14 ± 1 (6)	15 ± 0.5 (6)
ACE	1386 ± 48 (7)	1347 ± 60 (7) [a]

[a] ADX vs. sham-ADX.   [b] $P < 0.01$ by t-test

A very recent collaborative study with the laboratory of Dr. Paul Greengard has provided us with the neuronal counterpart of GPDH in the studies cited above, namely protein I. Protein I is a phosphoprotein constituent of synaptic vesicles which is implicated in synaptic physiology (see Nestler et al. 1981a, b). Corticosterone treatment of adrenalectomized rats for 24 h up to 2 weeks increases radioimmunoassayable protein I in hippocampus (Nestler et al. 1981a, b). The increase occurs in all hippocampal subfields known to have large numbers of glucocorticoid receptors and is not found in a variety of other brain regions with lesser glucocorticoid receptor levels. Finally, protein I levels do not increase after 1 week of exposure to 25% dexamethasone pellets, a finding which makes this effect clearly distinct from the GPDH elevation in hippocampus and from the pituitary ACTH reduction (Table 4).

Table 4. Glucocorticoid regulation of neuronal and glial properties in hippocampus and pituitary ACTH levels. Italizised values are significantly different from control ADX $P < 0.05$ or less Student's t test

Treatment	Glial GPDH [a]	Neuronal protein I [b]	Pituitary ACTH [c]
Control ADX	410 ± 27	56.8 ± 0.9	1.33 ± 0.09
+ 100% CORT	*628 ± 27*	*74.2 ± 2.3*	1.24 ± 0.17
+ 25% DEX	*594 ± 54*	56.4 ± 1.1	*0.87 ± 0.17*

[a] Meyer et al. 1979 μ mols/g protein/h
[b] Nestler et al. 1981a, b   nmols/g protein
[c] Gibson et al. 1979 μg/aliquot

There are several possible interpretations of an increase in the level of a prominent neuronal constituent such as protein I. On the one hand, it may result from increased growth of terminal fields. There is evidence against this, however, in that glucocorticoids *reduce* lesion-induced axon sprouting in the dentate gyrus (Scheff et al. 1980). On the other hand, changes in protein I may result from increased production of neurotransmitter candidates stored in vesicles containing protein I. Which transmitter? It would seem from the lack of glucocorticoid effects on enzymes of GABA and glutamate as well as acetylcholine metabolism (Table 3) that these three transmitter systems are not likely candidates. Regarding the neuropeptides, it has been found that glucocorticoids in the drinking water elevates levels of vasoactive intestinal polypeptide (VIP) in the hippocampus of adrenalectomized rats (Rotszteijn et al. 1980). Since VIP-containing cell bodies are present in subiculum, Ammon's horn, and dentate gyrus (Loren et al. 1979), this finding may have some relevance to the protein I story. However, there is one inconsistency, in that Rotsztejn et al. (1980) find that dexamethasone as well as corticosterone elevates VIP levels, whereas protein I is unaffected by dexamethasone (Table 4).

Another action of corticosterone on the hippocampus is to alter high affinity GABA uptake (Miller et al. 1978). Adrenalectomy leads within 24 h to an increase in GABA uptake into synaptosomes of 30%–40%. Corticosterone replacement therapy by the solid pellet method reverses the increase within 4–7 days (Miller et al. 1978). Corticosterone effects on GABA uptake are found in hippocampus but not in cerebral cortex and cerebellum (Table 5).

The direction of the effects of ADX on GABA uptake (to increase it) would be expected to remove extracellular GABA and thus reduce the inhibitory effect of this transmitter on hippocampal function. Increased electrical excitability and paroxysmal activity might result. This is what was reported for adrenalectomy in rats (Feldman and Robinson 1968). Such changes in GABA uptake and electrical activity may be related to increased susceptibility to audiogenic seizures and decreased electroconvulsive shock threshold with ADX, as reported by Woodbury (1954).

Table 5. The tissue specificity of the effect of adrenalectomy on the Vmax of GABA uptake. [a]

Tissue	Vmax (pmol/min/μg)	
	Normal	ADX
Hippocampus	0.99 ± 0.04	1.33 ± 0.01
Cerebellum	0.81 ± 0.01	0.81 ± 0.04
Frontal Cortex	1.18 ± 0.14	1.17 ± 0.09

[a] Synaptsomes were prepared from the tissue indicated. Values of Vmax were determined by double-reciprocal analysis. Values are the mean ± SEM of at least two independent determinations. From Miller et al. (1978)

The direction of corticosterone effects on GABA uptake (i.e., to decrease it) might be expected to potentiate the inhibitory effects of GABA on neurotransmission within the hippocampus. This prediction is consistent with observations of decreased single unit activity within Ammon's horn following corticosterone infusion (Pfaff et al. 1971). However, the time course of such effects on firing rate is considerably faster than for the effects on GABA uptake. There are two biochemical effects in hippocampus with a time course consistent with the rapid effects of corticosterone on firing rate of hippocampal neurons, namely, stimulation of the labeling of RNA in cell nuclei of hippocampus (Dokas 1979) and stimulation of the labeling of a protein in the soluble portion of hippocampal tissue (Etgen et al. 1979, 1980). Thus, the electrical effects of corticosterone might conceivably arise through a genomic action of the hormone. This relationship requires further study.

4 Behavioral and Neuroendocrine Aspects of Corticosterone Action in Hippocampus

4.1 Background

One goal of studying corticosterone action on the hippocampus is to understand what influence this interaction has on behavior and neuroendocrine function. Before such studies can begin, we must know what role the hippocampus plays in brain function. This question has occupied anatomists, psychologists, and neuroendocrinologists for many years (Nauta 1963; Douglas 1967; Kimble 1968; van Hartesveldt 1975; O'Keefe and Nadel 1978) and a review of such an immense topic is far beyond the scope of this chapter. In fact, there are many views as to the role of the hippocampus and as a result there has been very little systematic investigation of what glucocorticoids do to influence its role in the brain. Recently, however, certain hippocampal functions have begun to emerge as having relevance for glucocorticoid feedback and these will be summarized here, together with the research strategies from which they are derived.

4.2 Behavior

One research strategy was suggested to us (McEwen et al. 1975) by the extensive literature regarding the effects of hippocampal lesions on behavior (e.g., Douglas 1967; Kimble 1968). It was clear from these studies that only certain behaviors are disrupted by hippocampal damage, and we reasoned that these behaviors are the ones to study in the search for glucocorticoid influences. We initially studied three behaviors which are affected by hippocampal damage: spontaneous alternation, locomotor activity, and extinction of an appetitive runway task. The first two were unaffected in nondamaged adrenalectomized rats by corticosterone administration (Micco and McEwen 1980). However, for extinction of the runway task, adrenalectomy had the effect of facilitating extinction without altering acquisition; moreover, the corticosterone replacement in ADX rats retarded the rate of extinction and returned it to normal (Fig. 7); Micco et al. 1979; Micco and McEwen 1980). In keeping with the selectivity of hippocampal neuronal uptake of glucocorticoids, corticosterone replacement was effective but dexamethasone administration was not (Micco and McEwen 1980).

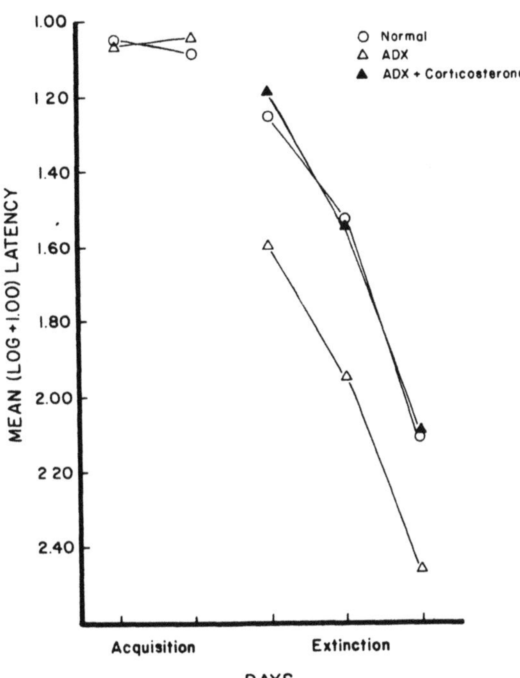

Fig. 7. Mean logarithmic running speeds for final 2 acquisition days and 3 extinction days. From Micco et al. (1979)

The hormone susceptibility of extinction, as opposed to the other two behaviors, lacks a firm explanation. We have suggested that a particular level of behavioral arousal may be needed to invoke the hormone-sensitive neurons (see Micco et al. 1979; Micco and McEwen 1980).

A main finding of these studies of runway extinction is that the direction of corticosterone action is the same as that of hippocampal ablation, as if corticosterone were inhibiting hippocampal function. Another recent finding (Micco et al. 1980) has reinforced this notion. ADX rats with electrodes in hippocampus were monitored by an automated recorder during 12 h light and 12 h dark periods in an isolated chamber. Activity, REM sleep, and slow-wave sleep plus quiet awake states were scored in 3-h segments. The effects of corticosterone replacement on the temporal distribution of these behavioral states revealed a marked increase in activity in the 3 h preceding light onset (Fig. 8). This result is noteworthy because another study has shown increased activity at the same time of day in rats with dorsal hippocampal ablation (Iuvone and van Hartesveldt 1977).

The idea that glucocorticoids inhibit hippocampal influences on behavior and brain function is consistent with other findings such as the rapid inhibition of hippocampal electrical activity by corticosterone (Pfaff et al. 1971) and the slower inhibition of high affinity GABA reuptake within hippocampus (see earlier discussion). One possible result of such inhibition might be to retard the action of the hippocampus in analyzing incoming sensory information (Nauta 1963). A role of the hippocampus in selective attention, which would contribute to the animal's attending to behaviorally relevant stimuli, has been suggested on the basis of lesion studies (Douglas 1967), and has also emerged from an electrophysiologic analysis of the influence of the nucleus locus ceruleus on hippocampal pyramidal cell electrical activity (Segal and Bloom 1976).

It is noteworthy in this connection that the removal of locus ceruleus input to hippocampus and cerebral cortex, by lesions of the dorsal noradrenergic bundle, results in increased distractability during the performance of previously trained reponses (Roberts et al. 1976). Moreover, a deficit in selective attention to relevant stimuli has been found in other experiments (Mason and Lin 1980) and such a deficit has been offered as a possible explanation for the resistance to extinction of rats with dorsal bundle lesions (Mason and Iversen 1979). The resistance to extinction after dorsal bundle lesions is similar in direction for at least one behavioral paradigm, an appetitive runway task, to the effects of hippocampal ablation (Mason and Iversen 1975; Jarrard et al. 1964; Raphaelson et al. 1966; Micco et al. 1979). The fact that in the same paradigm adrenalectomy facilitates and corticosterone retards extinction in nonlesioned rats (Micco et al. 1979) suggests an influence of adrenocortical hormones which is opposite to the normal action of the dorsal bundle, i.e., glucocorticoids may suppress mechanisms affecting selective attention which are normally facilitated by dorsal bundle input (McEwen and Micco 1980).

Through speculative, this idea has some indirect support. First, bilateral adrenalectomy abolishes the extinction-retarding effect of dorsal bundle lesions on an appetitive task (Mason et al. 1979). Secondly, ADX potentiates the increase in number of putative beta adrenergic receptors in hippocampus which results from dorsal bundle lesions (Roberts and Bloom 1980). In this connection, ADX also increases the noradrenalin-activated adenylate cyclase in rat cerebral cortex (Mobley and Sulser 1980), and this observation serves as a reminder that cerebral cortex may be involved both as a glucocorticoid target as well as an afferent neural target of the dorsal bundle projection. It is therefore a mistake to expect identical effects of hippocampal ablation and dorsal bundle lesions for every behavioral situation. It is also somewhat premature to equate behavioral effects of glucocorticoids with glucocorticoid actions on the hippocampus, although the results which have been summarized in this section offer real promise of future progress.

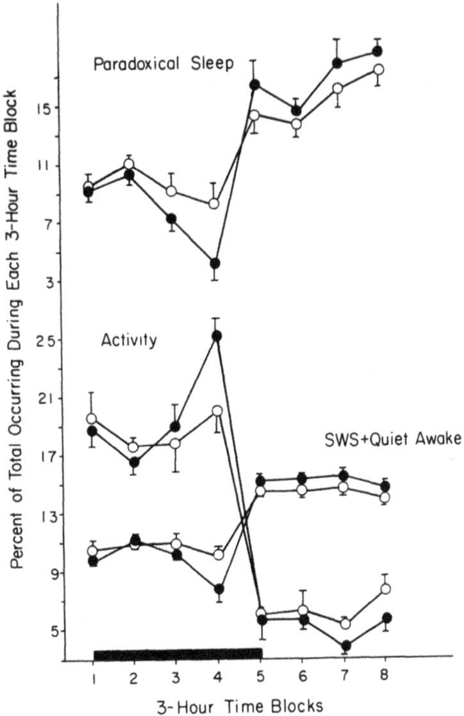

Fig. 8. Summary of the temporal distribution of three categories of behavior recorded automatically as described in the text. $\circ$, ADX; $\bullet$, ADX + CORT. *Ordinate,* percentage of total 24 h time in each category occurring during each 3-h time block (mean ± SEM). *Abscissa,* 3-h time block. *Dark bar* indicates lights-off period. Note that the ordinate value for each time block is presented at the beginning of the time block and represents the percentage of total 24 h time occurring in the 3 h which follows. Thus, for example, the marked decrease in activity recorded in time *block 5* reflects the decrease in activity which results from turning the lights on. Analysis of variance for each category (all $P < 0.01$): PS, ADX – $F_{7,40} = 7.50$; PS, ADX + CORT – $F_{7,40} = 19.51$; activity, ADX – $F_{7,40} = 23.65$; activity, ADX + CORT – $F_{7,40} = 50.16$; SWS + quiet awake, ADX – $F_{7,40} = 16.08$; SWS + quiet awake, ADX + CORT – $F_{7,40} = 44.48$. From Micco et al. (1980)

4.3 Neuroendocrine Control of Pituitary Adrenal Function

The principle object of neuroendocrine investigation of the hippocampus has been the pituitary—adrenal axis. There is no agreement in the literature as to whether the hippocampus has a facilitatory or inhibitory influence on ACTH secretion, owing to conflicts between results of electrical stimulation and lesion experiments as well as variation of reported effects depending on the time of day of the experiment (McEwen et al. 1972b; van Hartesveldt 1975). As to the role of glucocorticoid feedback in hippocampus on ACTH secretion, there are a few interesting results which can be summarized, and these lead to interesting speculations.

First, Slusher (1966) reported that implants of cortisone in the hippocampus (as well as in midbrain) caused the diurnal rhythm of corticosterone secretion to level off, with higher morning values and lower levels in the afternoon. In this connection, bilateral section of the fornix (Moberg et al. 1971), and in particular the medial cortico-hypo-thalamic tract (Fischette et al. 1980), disrupts the diurnal rhythm of glucocorticoid secretion. One of these studies (Fischette et al. 1980) suggests that hippocampal output may be primarily inhibitory, thus accounting for lower corticosterone levels in the morning in nonlesioned animals. If this were the case, at least in the morning, then the effects of cortisone implants (Slusher 1966) could be viewed as counteracting the normal hippocampal influence. A possible parallel to this inhibition by hippocampus of ACTH secretion exists in the aforementioned actions of the hippocampus in the morning to retard locomotor activity (Iuvone and van Hartesveldt 1977) and the apparent ability of corticosterone to block this inhibition and thereby mimic the effects of hippocampal ablation (Fig. 8; Micco et al. 198). (See discussion above.)

There is evidence that in the afternoon the hippocampus may retard ACTH secretion, insofar as hippocampal ablation is reported to result in higher basal and 2-min-ether-stress-induced ACTH concentrations (Wilson et al. 1980). Adrenalectomy abolished the difference in afternoon ACTH concentrations between hippocampal-lesioned and control cortical-lesioned rats. This finding suggests that corticosterone in the afternoon may be potentiating, rather than inhibiting, the postulated inhibitory hippocampal influence on ACTH secretion. To make these observations compatible with those described in the previous paragraph one would have to suppose that the corticosterne feedback might be inhibitory in the morning and facilitatory in the afternoon. It is noteworthy that in the study of Wilson et al. (1980) there was no indication of an effect of hippocampal ablation in the morning, in contrast to the results described in the paragraph above. Thus, discrepancies in the results of two related types of lesions (fornix versus hippocampus) complicate for the present our understanding of the action of the hippocampus on ACTH secretion.

Another aspect of the hippocampus and glucocorticoid feedback deals with the ability of dorsal hippocampal lesions and fornix transection to reduce the susceptibility of the animal to the negative feedback actions of dexamethasone on adrenocortical activity (Wilson 1975; Feldman and Conforti 1976, 1980). Although these results may indicate the participation of the hippocampus as a target for dexamethasone feedback, it is also possible that ". . . dexamethasone may exert its influence on the pituitary, and the effect observed may be the summation of dexamethasone action on the pituitary gland plus lack of hippocampal input" (Feldman and Conforti 1980). This statement may be equally applicable as a cautionary note to the other instances where glucocorticoid feedback effects have been found to differ between brain-lesioned and nonlesioned animals, in the sense that lesions may alter the sensitivity of other neural structures or the pituitary to hormonal feedback.

5 Conclusions

The discovery of glucocorticoid receptors in rat hippocampus provided the impetus for studies of this structure which have revealed hormone-induced increases in RNA and protein synthesis, decreases in neuronal electrical activity, decreases in high affinity synaptosomal reuptake of GABA, and increases in the level of a neuronal synaptic vesicle phosphoprotein. It is likely that this list will continue to grow and that the items on it will eventually provide an explanation for the hormone-induced alterations of hippocampal function. As noted earlier, for example, it is possible to see how the increased uptake of GABA which follows adrenalectomy may help explain the enhanced seizure susceptibility of the brain in the ADX state.

It is also evident that glial cells in hippocampus and in other neural structures contain glucocorticoid receptors and that these receptors interact with dexamethasone as well as with corticosterone. The most prominent effect of glucocorticoids on glial cells is the induction of glycerol phosphate dehydrogenase in oligodendoglial cells and G6 glial tumor cells (see above discussion). Glucocorticoids also inhibit myelination (Friedrich and Bohn 1980) which is a function of the oligodendoglial-like cells, and they promote hypertrophy of astrocytes (Scheff et al. 1980). It remains to be established how glucocorticoid action on glial cells of adult animals influence neuronal properties and neural activity. It is conceivable, for example, that changes in glial cells may account for altered nerve conduction velocities found as a result of adrenocortical hyper- and hyposecretion (Henkin 1970).

The discovery that the neurons of the hippocampus of rodents and primates (and possibly of all mammals) contain putative receptors for corticosterone has added a new dimension to the puzzle of the hippocampus and its role in brain function and behavior. Certain instances in which corticosterone treatment produces effects in the same direction as those of hippocampal ablation suggested to us as a working hypothesis that glucocorticoid action may suppress aspects of hippocampal function. This hypothesis is supported by both electrophysiologic and neurochemical studies on the influence of corticosterone on hippocampal neurons.

The hypothesis takes a further, far more speculative, form with the suggestion that corticosterone may act to suppress the role of the hippocampus in filtering out behaviorally irrelevant sensory stimuli (McEwen and Micco 1981). The greater distractibility of the animal which might result from such corticosterone action could help explain the reduced rate of extinction of food-seeking behavior (Micco et al. 1979). Possible variation in distractibility with adrenal hormone excess or insufficiency could be tested experimentally in both rodents and in human patients. In this connection, it is worthwhile noting that adrenocorticoid insufficiency in humans has been associated with decreases in thresholds for detection of sensory cues and increases in thresholds for sensory recognition (Henkin 1970). The relationship of these effects to postulated changes in distractibility remains to be established.

Acknowledgments. Research in the author's laboratory on this topic was supported by research grant NS07080 to the author from the USPHS and by Grant GB43558 to the author and Dr. David J. Micco, jr. from the NSF. The editorial assistance of Mrs. Oksana Wengerchuk is gratefully acknowledged.

References

Bush IE (1953) Species differences in adrenocortical secretion. J Endocrinol 9:95–100

De Kloet ER, McEwen BS (1976) A putative glucocorticoid receptor and a transcortin-like macromolecule in pituitary cytosol. Biochim Biophys Acta 421:115–123

De Kloet ER, Wallach G, McEwen BS (1975) Differences in corticosterone and dexamethasone binding to rat brain and pituitary. Endocrinology 96:598–609

De Kloet ER Burbach P, Mulder GH (1977) Localization and role of transcortin-like molecules in the anterior pituitary. Mol Cell Endocrinol 7:261–273

DeVellis J, Inglish D (1968) Hormonal control of gylcerol phosphate dehydrogenase in the rat brain. J Neurochem 15:1061–1070

DeVellis J, Inglish D (1973) Age-dependent changes in the regulation of glycerol-phosphate dehydrogenase in the rat brain and in a glial cell line. Prog Brain Res 40:322–330

DeVellis J, Inglish D, Cole R, Molson J (1971) Effects of hormones on the differentiation of cloned lines of neurons and glial cells. In: Ford D (ed) Influence of hormones on the nervous system. Karger, Basel Paris London New York, pp 25–39

DeVellis J, McEwen BS, Cole R, Inglish D (1974) Relations between glucocorticoid nuclear binding, cytosol recpeotr activity and enzyme induction in a rat glial cell line. J Steroid Biochem 5:392–393

Dokas LA (1979) Corticosterone and RNA metabolism in the rat hippocampus (Abstr 1498). Soc Neurosci Ann Mtg Atlanta GA, p 443

Douglas RJ (1967) The hippocampus and behavior. Psychol Bull 67:416–442

Etgen AM, Lee KS, Lynch G (1979) Glucocorticoid modulation in specific protein metabolism in hippocampal slices maintained in vitro. Brain Res 165:37–45

Etgen AM, Martin M, Gilbert R, Lynch G (1980) Characterization of corticosterone-induced protein synthesis in hippocampal slices. J Neurochem 35:598–602

Feldman S, Conforti N (1976) Feedback effects of dexamethasone on adrenocortical responses in rats with fornix section. Horm Res 7:56–60

Feldman S, Conforti N (1980) Participation of the dorsal hippocampus in the glucocorticoid feedback effect on adrenocortical activity. Neuroendocrinology 30:52–55

Feldman S, Robinson S (1968) Electrical activity of the brain in adrenalectomized rats with implanted electrodes. J Neurol Sci 6:1–8

Fischette CT, Komisaruk BR, Edinger HM, Feder HH, Siegel A (1980) Differential fornix ablations and the circadian rhythmicity of adrenal corticosteroid secretion. Brain Res 195:373–387

Friedrich VL Jr, Bohn MC (1980) Glucocorticoids inhibit myelination in developing rat (Abstr 1323). Soc Neurosci Ann Mtg, Cincinnati OH, p 380

Gerlach JL, McEwen BS (1972) Rat brain binds adrenal steroid hormone: radioautography of hippocampus with corticosterone. Science 175:1133–1136

Gerlach JL, McEwen BS, Pfaff DW, Moskovitz S, Ferin M, Carmel PW, Zimmerman EA (1976) Cells in regions of rhesus monkey brain and pituitary retain radioactive estradiol, corticosterone and cortisol differentially. Brain Res 103:603–612

Gibson MJ, Krieger DT, Liotta AS, Brownstein MJ, McEwen BS (1979) Chronic dexamethasone alters content of immunoreactive ACTH-like material in rat arcuate nucleus (Abstr 1507). Soc for Neurosci Ann Mtg, Atlanta GA, p 446

Grosser BI, Stevens W, Bruenger FW, Reed DJ (1971) Corticosterone binding in rat brain cytosol. J Neurochem 18:1725–1732

Grosser BI, Stevens W, Reed DJ (1973) Properties of corticosterone-binding macromolecules from rat brain cytosol. Brain Res 57:387–395

Hartesveld C van (1975) The hippocampus and regulation of the hypothalamic-hypophysial-adrenal cortical axis. In: Isaacson RL, Pribram KH (eds) The hippocampus: a comprehensive treatise. Plenum Press, New York, pp 375–391

Henkin RI (1970) The neuroendocrine control of perception. In: Hamburg DA, Pribram KH, Stunkard AS (eds) Perception and its disorders. Williams and Wilkins, Baltimore (Research publication of the association for research in nervous and mental disease, vol XLVII, pp 54–107)

Iuvone PM, van Hartesveldt C (1977) Diurnal locomotor activity in rats: effects of hippocampal ablation and adrenalectomy. Behav Biol 19:228–237

Jarrard LE, Isaacson RL, Wickelgren O (1964) Effects of hippocampal ablation and intertrial interval on runway acquisition and extinction. J Comp Physiol Psychol 57:442–444

Jensen EV, Suzuki T, Kawashima T, Stumpf WE, Jungblut PW, De Sombre ER (1968) A two-step mechanism for the interaction of estradiol with rat uterus. Proc Nat Acad Sci USA 59:632–638

Kimble DP (1968) Hippocampus and internal inhibition. Psychol Bull 70:285–295

Knizley H Jr (1972) The hippocampus and septal area as primary target sites for corticosterone. J Neurochem 19:2737–2745

Koch B, Lutz B, Briaud B, Mialhe C (1975) Glucocorticoid binding to adenohypophysis receptors and its physiological role. Neuroendocrinology 18:299–310

Koch B, Lutz-Bucher B, Briaud B, Mialhe C (1977) Glucocorticoid binding to plasma membranes of the adenohypophysis. J Endocrinol 73:399–400

Leveille PJ, McGinnis JF, Maxwell DS, DeVellis J (1980) Immunocytochemical localization of glycerol-3-phosphate dehydrogenase in rat oligodendrocytes. Brain Res 196:287–305

Loren I, Alumets J, Hakanson R, Sundler F (1979) Distribution of gastrin and CCK-like peptides in rat brain. Histochemistry 59:249–257

MacLusky NJ, Chaptal C, Lieberburg I, McEwen BS (1976) Properties and subcellular inter-relationships of presumptive estrogen receptor macromolecules in the brains of neonatal and prepubertal female rats. Brain Res 114:158–165

MacLusky NJ, Turner BB, McEwen BS (1977) Corticosteroid binding in rat brain and pituitary cytosols: resolution of multiple binding components by polyacrylamide gel based isoelectric focusing. Brain Res 130:564–571

Mason ST, Iversen SD (1975) Learning in the absence of forebrain noradrenaline. Nature 258:422–424

Mason ST, Iversen SD (1979) Theories of the dorsal bundle extinction effect. Brain Res 1:107–137

Mason ST, Lin D (1980) Dorsal noradrenergic bundle and selective attention in the rat. J Comp Physiol Psychol 94:819–832

Mason ST, Roberts DCS, Fibiger HC (1979) Interaction of brain noradrenaline and the pituitary-adrenal axis in learning and extinction. Pharmacol Biochem Behav 10:11–16

McEwen BS, Micco DJ Jr (1980) Toward an understanding of the multiplicity of glucocorticoid actions on brain function and behavior. In: van Keep PA, DeWied D (eds) The brain as an endocrine target organ in health and disease. MTP Press, Lancaster, pp 11–28

McEwen BS, Pfaff DW (1973) Chemical and physiological approaches to neuroendocrine mechanisms: attempts at integration. In: Ganong WF, Martini L (eds) Frontiers in neuroendocrinology. Oxford University Press, New York, pp 267–335

McEwen BS, Plapinger L (1970) Association of corticosterone-1,2-H^3 with macromolecules extracted from brain cell nuclei. Nature 226:263–264

McEwen BS, Wallach G (1973) Corticosterone binding to hippocampus: nuclear and cytosol binding in vitro. Brain Res 57:373–386

McEwen BS, Weiss JM, Schwartz L (1968) Selective retention of corticosterone by limbic structures in rat brain. Nature 220:911–912

McEwen BS, Weiss JM, Schwartz LS (1969) Uptake of corticosterone by rat brain and its concentration by certain limbic structures. Brain Res 16:227–241

McEwen BS, Weiss JM, Schwartz LS (1970) Retention of corticosterone by cell nuclei from brain region of adrenalectomized rats. Brain Res 17:471–482

McEwen BS, Magnus C, Wallach G (1972a) Soluble corticosterone-binding macromolecules extracted from rat brain. Endocrinology 90:217–226

McEwen BS, Zigmond RE, Gerlach JL (1972b) Sites of steroid binding and action in the brain. In: Bourne GH (ed) Structure and function of nervous tissue, vol 5. Academic Press, New York, pp 205–291

McEwen BS, Gerlach JL, Micco DJ Jr (1975) Putative glucocorticoid receptors in hippocampus and other regions of the rat brain. In: Isaacson R, Pribram K (eds) The hippocampus: a comprehensive treatise. Plenum Press, New York, pp 285–322

McEwen BS, deKloet R, Wallach G (1976) Interactions in vivo and in vitro of corticoids and progesterone with cell nuclei and soluble macromolecules from rat brain regions and pituitary. Brain Res 105:129–136

Meyer JS, Micco DJ, Stephenson BS, Krey LC, McEwen BS (1979a) Subcutaneous implantation method for chronic glucocorticoid replacement therapy. Physiol Behav 22:867–870

Meyer JS, Luine VN, Khylchevskaya RI, McEwen BS (1979b) Glucocorticoids and hippocampal enzyme activity. Brain Res 166:172–175

Meyer JS, McEwen BS (1982a) Evidence for glucocorticoid target cells in the rat optic nerve. Physicochemical characterization of cytosol binding sites. J Neurochem (in press)

Meyer JS, Leveille PJ, deVellis J, Gerlach JL, McEwen BS (1982b) Evidence for glucocorticoid target cells in the rat optic nerve. Hormone binding and glycerolphosphate dehydrogenase induction. J Neurochem (in press)

Micco DJ Jr, McEwen BS (1980) Glucocorticoids, the hippocampus and behavior: interactive relation between task activation and steroid hormone binding specificity. J Comp Physiol Psychol 94:624–633

Micco DJ, McEwen BS, Shein W (1979) Modulation of behavioral inhibition in appetitive extinction following manipulation of adrenal steroids in rats: implications for involvement of the hippocampus. J Comp Physiol Psychol 93:323–329

Micco DJ Jr, Meyer JS, McEwen BS (1980) Effects of corticosterone replacement on the temporal patterning of activity and sleep in adrenalectomized rats. Brain Res 200:206–212

Miller AL, Chaptal C, McEwen BS, Peck EJ Jr (1978) Modulation of high affinity GABA uptake into hippocampal synaptosomes by glucocorticoids. Psychoneuroendocrinology 3:155–164

Moberg GP, Scapagnini V, DeGroot J, Ganong WF (1971) Effect of sectioning the fornix on diurnal fluctuation in plasma corticosterone levels in the rat. Neuroendocrinology 7:11–15

Mobley PL, Sulser F (1980) Adrenal corticoids regulate sensitivity of noradrenaline receptor-coupled adenylate cyclase in brain. Nature 286:608–609

Nauta WJH (1963) Central nervous organization and the endocrine motor system. In: Nalbandor AV (ed) Advances in neuroendocrinology. University of Illinois Press, Urbana

Nestler EJ, Rainbow TC, McEwen BS, McEwen P (1981a) Corticosterone increases the level of protein I, a neuron-specific protein, in rat hippocampus. Science 212: 1162–1164

Nestler EJ, Rainbow TC, McEwen BS, Greengard P (1981b) Effect of steroid hormones on the level of protein I in rat brain. In: Fuxe K et al. (eds) Steroid hormone regulation of the brain. Pergamon Press, Oxford New York, pp 205–216

O'Keefe J, Nadel L (1978) The hippocampus as a cognitive map. Clarendon Press, Oxford

Olpe H-R, McEwen BS (1976) Glucocorticoid binding to receptor-like proteins in rat brain and pituitary: ontogenetic and experimentally induced changes. Brain Res 105:121–128

Pfaff DW, Silva MTA, Weiss JM (1971) Telemetered recording of hormone effects on hippocampal neurons. Science 172:394–395

Raphelson AC, Isaacson RL, Douglas RJ (1966) The effect of limbic damage on the retention and performance of a runway response. Neuropsychologia 4:253–264

Rees HD, Stumpf WE, Sar M (1975) Autoradiographic studies with ^{3}H-dexamethasone in the rat brain and pituitary. In: Stumpf WE, Grant LA (eds) Anatomical neuroendocrinology. Karger, Basel, pp 262–269

Rhees RW, Grosser BI, Stevens W (1975a) Effect of steroid competition and time on the uptake of (^{3}H) corticosterone in the rat brain; an autoradiographic study. Brain Res 83:293–300

Rhees RW, Grosser BI, Stevens W (1975b) The autoradiographic localization of ^{3}H-dexamethasone in the brain and pituitary of the rat. Brain Res 100:151–156

Roberts DCS, Bloom FE (1980) Adrenal influence on noradrenaline receptors in the hippocampus (Abstr 151.23). Soc Neurosci Ann Mtg, Cincinnati OH, p 446

Roberts DCS, Price MTC, Fibiger HC (1976) The dorsal tegmental noradrenergic projection: an analysis of its role in maze learning. J Comp Physiol Psychol 90: 363–372

Rotszteijn WH, Normand M, Lalonde J, Fortier C (1975) Relationship between ACTH release and corticosterone binding by the receptor sites of the adenohypophysis and dorsal hippocampus following infusion of corticosterone at a constant rate in the adrenalectomized rat. Endocrinology 97:223–230

Rotszteijn WH, Besson J, Briaud B, Gagnant L, Rosselin G, Kordon C (1980) Effect of steroids on vasoactive intestinal peptide (VIP) in discrete brain regions and peripheral tissues. Neuroendocrinology 31:287–291

Scheff SW, Bernado LS, Cotman CW (1980) Hydrocortisone administration retards axon sprouting in the rat dentate gyrus. Exp Neurol 68:195–201

Segal M, Bloom FE (1976) The action of norepinephrine in the rat hippocampus. IV. The effects of locus coerulus stimulation on evoked hippocampal unit activity. Brain Res 107:513–525

Slusher MA (1966) Effects of cortisol implants in the brainstem and ventral hippocampus on diurnal corticosterone levels. Exp Brain Res 1:184–194

Stevens W, Grosser BI, Reed DJ (1971) Corticosterone-binding molecules in rat brain cytosols: regional distribution. Brain Res 35:602–607

Stumpf WE (1971) Autoradiographic techniques and the localization of estrogen, androgen, and glucocorticoid in the pituitary and brain. Am Zool 11:725–739

Turner BB, McEwen BS (1980) Hippocampal cytosol binding capacity of corticosterone: no depletion with nuclear loading. Brain Res 189:169–182

Warembourg M (1975a) Radioautographic study of the rat brain after injection of (1,2,^{3}H) corticosterone. Brain Res 89:61–70

Warembourg M (1975b) Radioautographic study of the rat brain and pituitary after injection of ^{3}H dexamethasone. Cell Tissue Res 161:183–191

Wilson MM (1975) Effect of hippocampectomy on dexamethasone suppresion of corticosteroid-sensitive stress responses (Abstr 511) Anat Rec, p 181

Wilson MM, Greer SE, Greer MA, Roberts L (1980) Hippocampal inhibition of pituitary-adrenocortical function in female rats. Brain Res 197:433–441
Woodbury DM (1954) Effects of hormones on brain excitability and electrolytes. Recent Prog Horm Res 10:65–107
Wrange O (1979) A comparison of the glucocorticoid receptor in cytosol from rat liver and hippocampus. Biochim Biophys Acta 582:346–357

Glucocorticoids and the Developing Nervous System

ALLISON J. DOUPE and PAUL H. PATTERSON[1]

Contents

Abbreviations

ACTH: adrenocorticotropic hormone; DβH: dopamine-β-hydroxylase; EGF: epidermal growth factor; GPDH: glycerol phosphate dehydrogenase; HIOMT: hydroxyindole-O-methyltransferase; NGF: nerve growth factor; PNMT: phenylethanolamine-N-methyl-transferase; TH: tyrosine hydroxylase; TrpH: tryptophan hydroxylase

1 Introduction

The environment of a developing neuron is rich in signals which can influence its fate. Some striking examples of this are provided by neural crest cells whose destination and phenotype may be influenced by the environment through which they migrate (Cohen 1972; Teillet et al. 1978), by possible chemotactic factors such as nerve growth factor (e.g., Gundersen and Barrett 1979), by factors produced by target tissues (Patterson and Chun 1977a; Teillet et al. 1978), and by electrical activity (Walicke et al. 1977). Glucocorticoid hormones are another likely influence on developing neurons: they have a variety of actions on the adult brain (McEwen 1978; McEwen et al. 1979), their synthesis begins during prenatal life (Jost 1966), and they are known to have numerous effects on differentiation in many tissues (Ballard 1979). A common action of cortico-steroids is the acceleration of specific developmental events. For instance, the striking

1 Department of Neurobiology, Harvard Medical School, Boston, MA 02115, USA

increase in serum corticoid level which occurs at term as a result of increased fetal adrenal activity (e.g., Martin et al. 1977; Mulay et al. 1973) causes a wide variety of maturational changes as a "preparation for birth" (Liggins 1976). Postnatally, glucocorticoid levels fall sharply, then rise again, assuming a diurnal rhythm, as the hypothalamic-pituitary-adrenal axis matures (Allen and Kendall 1967; Ramaley 1974). This rise, which elicits further differentiation, occurs at about postnatal day 19 (P19) in the rat.

Like other steroids, glucocorticoids are thought to act by binding to cytoplasmic receptors and moving to the nucleus, where they may regulate transcription (see Johnson et al. 1979). Although this sequence has not yet been clearly demonstrated for most actions of glucocorticoids in the adult nervous system, specific (and predominantly neuronal) corticosteroid binding is found in many adult rat brain areas (McEwen et al. 1975; Warembourg 1975a, b). The ontogeny of glucocorticoid receptors has not been studied in detail, but in some brain areas, the receptors are known to increase during development (Olpe and McEwen 1976). In other areas they have been found to be present long before the period of normal differentiation and maximum inducibility of a glucocorticoid-related event (Koehler and Moscona 1975; Lippman et al. 1974).

Because of the widespread effects of glucocorticoids it is difficult to determine in intact animals whether the results observed in one tissue or cell type are the result of direct hormonal actions on that tissue. For example, a number of studies have shown that corticosteroid administration to newborn animals has detrimental effects on brain development. In general, both DNA content and brain weight are permanently reduced (15%–18%), probably due to an inhibition of cell proliferation during the period of administration (Cotterrell et al. 1972; Howard 1968). However, the decreased brain weight is always accompanied by a decrease in body weight as well, and early nutritional deprivation and body weight loss can have significant effects on brain development: DNA content, neuronal growth and synaptogenesis, and myelination can all be decreased (see Jacobson 1978). Although receptor localization studies provide a first clue that glucocorticoids may be acting directly on a specific tissue or cell type, further evidence can be obtained from local hormone implantation or culture studies. We will discuss the relatively few cases in which there is evidence for a direct action of corticosteroids on the developing nervous system.

2 Chromaffin and Chromaffin-Like Cells

The sympathoadrenal system provides a number of examples of glucocorticoid effects on development. Adrenergic neural crest derivatives are of several types: sympathetic principal neurons, adrenal medullary chromaffin cells, and small intensely fluorescent (SIF) cells. These cells all produce catecholamines, but differ in their morphology, fine structure, and type of catecholamine synthesized. Mature sympathetic neurons have 40–50 μm cell bodies and long processes with specializations for release of transmitter at sites adjacent to target organs; their amine-storing vesicles, which contain norepinephrine, are small (50 nm) and are concentrated in the processes (Burnstock and Costa 1975). On the other hand, adrenal chromaffin cells have few or no processes, and their small (10–20 μm) cell bodies are packed with large amine-storing vesicles (100–350 nm), resulting in an intense formaldehyde-induced fluorescence and positive staining in the chromaffin reaction. Like other endocrine cells, chromaffin cells release their transmitter (norepinephrine or epinephrine) directly into the circulation (Coupland 1965a, b). Finally, SIF cells have some properties of both neurons and adrenal chromaffin cells. Their small cell bodies have numerous amine-storing vesicles which fall into two size classes: type II SIF cells have vesicles similar in diameter to adrenal medullary cells (150–300 nm) whereas type I SIF cells have smaller vesicles (100 nm) (Lu et al. 1976). SIF cells sometimes have processes, usually short (30–40 μm), although they can be longer (Furness and Costa 1976). Some SIF cells in ganglia make efferent synapses on neurons while others are in close apposition to fenestrated blood vessels (see Taxi 1979). The morphological heterogeneity of these cells is paralleled by heterogeneity in the catecholamine stored: many cells appear to contain only dopamine (Björklund et al. 1970; Fuxe et al. 1971; Rybarczyk et al. 1976), while others store norepinephrine or epinephrine (Elfvin et al. 1975).

How do the differences between these closely related cell types arise during development? Glucocorticoids are an important environmental signal influencing the choice of phenotype. The level of this hormone surrounding adrenal medullary cells (10^{-6} M) is more than 100-fold higher than in the general circulation, because the medulla receives the adrenocortical venous effluent (Coupland 1975; Jones et al. 1977). The preferential localization of SIF cells around blood vessels may also reflect local differences in hormone concentration. An initial suggestion that corticosteroid levels might be important in the chromaffin cell-neuron decision came from experiments in which adrenal medullary tissue grew neuron-like processes when transplanted from its hormone-rich environment to the relatively corticosteroid-poor anterior eye chamber (Olson 1970). In cultures of dissociated rat adrenal chromaffin cells, neurite outgrowth was elicited by addition of the growth factor for sympathetic neurons, nerve growth factor (NGF), but simultaneous addition of dexamethasone completely inhibited the fiber outgrowth (Unsicker et al. 1978). Thus the chromaffin cell phenotype may result from corticosteroid suppression of NGF-induced neuronal differentiation. Further support for an interaction between these two factors was obtained by Aloe and Levi-Montalcini (1979). When they injected NGF into rats beginning at embryonic day 17 (E17) (the time of early migration of nearby neural crest cells into the adrenal cortex), the adrenal medulla was almost completely replaced by neurons. Thus high levels of NGF may override the stimulus directing neural crest cells to become chromaffin cells. An alternative interpretation – that NGF stimulates mitosis and growth of neuroblasts, while killing chromaffin cells – seems unlikely in view of the results obtained with dissociated chromaffin cells in culture.

There is another population of chromaffin cells found outside the adrenal gland, primarily clustered around abdominal blood vessels and in some sympathetic ganglia. These extra-adrenal chromaffin cells also respond to glucocorticoids. In the normal rat, they are maximally chromaffin-positive at birth and degenerate postnatally (Coupland 1965a; Lempinen 1964). Corticosteroid administration to newborn animals prevents this degeneration. Furthermore, the extra-adrenal chromaffin tissue shows hyperplasia, and chromaffin cells appear in abundance in structures such as the thoracic sympathetic glanglia, where normally only occasional chromaffin cells are seen

(Lempinen 1964). Lempinen hypothesized that the normal degeneration of extra-adrenal chromaffin tissue was caused by the fall in circulating corticosteroids seen shortly after birth, and that he prevented this by giving hydrocortisone. However, the appearance of new chromaffin cells and the hyperplasia of the para-aortic extra-adrenal chromaffin tissue are more difficult to explain. It is possible that glucocorticoids (1) induced mitosis in these cells, (2) increased catecholamine synthesis in pre-existing cells (so that they became chromaffin-positive), or (3) induced the differentiation of chromaffin cells from precursor cells.

The same questions arise for glucocorticoid actions on SIF cells. Eränkö and colleagues found that hormone treatment at birth increased the number of SIF cells in adult rat superior cervical ganglia up to ten-fold over untreated siblings (Eränkö and Eränkö 1972; Eränkö et al. 1972a). This effect was observed both in vivo and in organ culture. Hydrocortisone treatment also increased the fluorescence intensity of SIF cells , the number that were chromaffin-positive, and the size and number of their granular vesicles (Eränkö et al. 1972b).

Further analysis of these hormonal actions is being carried out in dissociated cell culture. SIF cells from the newborn rat superior cervical ganglion can be grown in the virtual absence of neurons and non-neuronal cells by the addition of dexamethasone and antimitotic agents and the omission of nerve growth factor (Doupe et al. 1980). Under these conditions glucocorticoids can induce the appearance of SIF cells in a population of undifferentiated cells. This finding supports one of the previously mentioned mechanisms for hormone-induced hyperplasia. Furthermore, like adrenal chromaffin cells, SIF cells can respond to NGF with outgrowth of neuritic processes. Some cells also thereby lose their intense fluorescence and acquire neuronal ultrastructure (Doupe et al. 1980). This suggests that the SIF cell phenotype may reflect a balance between the local levels of glucocorticoid and NGF. In addition, preliminary results show that, in the absence of NGF, SIF cell morphology depends on the glucocorticoid concentration: at low (10^{-8} M) hormone concentration, the cells have predominantly small granular vesicles as in type I cells, and with increased hormone levels (10^{-6} M) the vesicle size approaches that of the type II and adrenal medullary cells (Doupe AJ, Patterson PH, Landis SC, unpublished work). Thus SIF cells may represent an environmentally determined intermediate state between neurons and chromaffin cells. An unresolved question is whether these different phenotypes also represent stages in normal development. In this regard it is interesting that in culture undifferentiated cells can become SIF cells, and then be converted into neurons.

3   Catecholamine Synthetic Enzymes

Another characteristic of the chromaffin cell phenotype which glucocorticoids influence is the enzyme phenylethanolamine-N-methyltransferase (PNMT), which catalyzes the methylation of norepinephrine to epinephrine. Administration of adrenocorticotropic hormone (ACTH) or dexamethasone to adult rats causes no change in PNMT activity, but hypophysectomy reduces the enzyme activity by 80% (Wurtman and Axelrod 1965). This decrease can be prevented or reversed by ACTH or corticosteroid.

The maintenance of PNMT activity by dexamethasone does not depend on enhanced enzyme synthesis but rather on slowed degradation (Ciaranello 1978). It appears that glucocorticoids exert their effect through a corticosteroid-dependent, endogenous stabilizing factor which protects PNMT against thermal and tryptic denaturation. There is evidence that this factor may be S-adenosyl methionine, a methyl donor in the reaction catalyzed by PNMT (Ciaranello et al. 1978).

Do glucocorticoids play a similar role in maintenance or induction of medullary PNMT during development? A point that must be kept in mind is the considerable variation among species in the number and distribution of epinephrine- and norepinephrine-storing cells within the adult medulla. For instance, almost 100% of guinea pig medullary cells have PNMT and store epinephrine, in contrast to 80% in rats and humans and 60% in cats (Poherecky and Wurtman 1971). It is not clear what determines this variability. Although the other catecholamine synthetic enzymes tyrosine hydroxylase (TH) and dopamine-β-hydroxylase (DβH) appear in migrating neural crest cells as early as E11-13 in the rat, the appearance of PNMT is first detected immunohistochemically and biochemically at E17 (Teitelman et al. 1979; Verhofstad et al. 1979). PNMT is seen only in the precursor cells which are in direct contact with the adrenal gland. Steroid synthesis in the adrenal cortex rises sharply at E16.5, and hypophysectomy of embryos at E18 results in very low PNMT activity at birth (Margolis et al. 1966). Furthermore, there is a positive correlation in many mammalian species between the fraction of total catecholamines represented by epinephrine and the ratio of the sizes of cortex and medulla (Shepherd and West 1951).

These findings suggested that glucocorticoids might induce PNMT during development. The variability in percentage of PNMT-containing cells might then be explained by the fact that a portion of the blood supply of the medulla comes directly from the arterial side of the circulation, and is thus relatively corticosteroid-poor, while the rest is cortical venous blood, with very high steroid levels (Coupland 1975). Cells lying close to the arterial supply might become norepinephrine-storing and those close to the venous supply would be induced to store epinephrine. However, a detailed comparison of the intra-adrenal vascular pattern of each species with the number and distribution of its catecholamine-storing cells has not yet been made.

On the other hand, recent evidence has raised some doubts about the hypothesis that glucocorticoids induce the expression of PNMT during development (Bohn et al. 1980, 1981). The initial appearance of the enzyme (as determined immunohistochemically) is not prevented or delayed by embryonic hypophysectomy or treatment with inhibitors of adrenocortical function. Furthermore, chronic or acute treatment of mothers and/or embryos with glucocorticoids or ACTH does not result in precocious induction of the enzyme, i.e., before E17. However, cellular levels of PNMT (as indicated by the intensity of the immunofluorescence) are lower after embryonic hypophysectomy. This suggests that the initial induction of PNMT is independent of glucocorticoid, but that the ontogenetic increase in PNMT levels after initial expression has occurred does require intact adrenal function. Similar results are obtained with cultured adrenal glands: removal on E16 and organ culture without added ACTH prevents the 100-fold surge in corticosteroid synthesis normally seen between E16–18. However, PNMT activity appears at the usual time, and increases four-fold in 3 days (Brodsky et al. 1980). Also, culture of the adrenal gland with 10^{-5} M corticosteroid does not change the time of appearance of PNMT, but does increase PNMT activity three-fold after it appears. Immunotitration showed that the increased activity was due to an increase in the number of enzyme molecules, but it is not yet clear whether this is due to increased synthesis of the enzyme or decreased degradation, as in the adult.

However, the interpretation of these recent experiments is not completely straight-forward. A low level of glucocorticoid may persist after the various experimental procedures and might be enough to induce initial expression of the enzyme. Lack of precocious induction by corticosteroid may simply signify that the cells are not competent to be induced until E17. The latter point may be supported by the observation that implantation of corticosteroid agonists into mothers actually did increase the number (and not just the fluorescence intensity) of PNMT-containing cells at E18.5, although not at E17.5 (Bohn et al. 1981). One explanation of this result is that once cells become competent for induction, increased glucocorticoid levels can induce the expression of PNMT. On the other hand, the hormone may simply have increased the amount of PNMT in additional cells to a level detectable by immunohistochemistry. An important experiment would be to increase the glucocorticoid levels in the arterial and/or the venous blood of the medulla and determine the number and distribution of PNMT-containing cells. Even if glucocorticoids are unlikey candidates for the inducer of the initial expression of PNMT, evidence for an alternative possibility remains to be presented.

The role of corticosteroids in PNMT induction in SIF and extra-adrenal chromaffin cells could be somewhat different. Administration of glucocorticoids to newborn animals elicits the appearance of epinephrine in para-aortic bodies which normally store norepinephrine (Coupland and MacDougall 1966; Eränkö et al. 1966). In the rat superior cervical ganglion at birth, there are low levels of epinephrine which have been attributed to SIF cells. If hydrocortisone is administered to newborn rats, both ganglionic PNMT activity and epinephrine content are increased (Ciaranello et al. 1973; Koslow et al. 1975). This effect is only seen if hormone treatment is begun during the first few postnatal days. After withdrawal of the hormone, epinephrine levels fall. Recently, PNMT has been localized immunohistochemically in the ganglion in vivo and in organ culture. Although not normally detectable at any stage, enzyme staining can be induced in clusters of small cells by hormone treatment at birth (Black et al. 1980). These small cells are likely to be SIF cells, although intense catecholamine fluoresence and PNMT staining have not yet been demonstrated in the same or adjacent sections. Dexamethasone treatment of pregnant rats can elicit positively-staining cells in fetuses as early as E18.5. Thus glucocorticoids appear to be capable of inducing both the initial expression and the precocious appearance of PNMT in SIF cells, in contrast to the situation in the adrenal medulla. However, even though PNMT is not detectable immunohistochemically at birth in untreated ganglia, it is biochemically measurable. Glucocorticoid administration could simply be increasing PNMT levels to the threshold for immunofluorescence detectability, while the initial expression of the enzyme was induced by some other factor. On the other hand, the PNMT measured biochemically at birth could be the result of the high prenatal levels of endogenous corticosteroids. An important experiment in resolving this issue will be biochemical analysis of PNMT in fetal and neonatal animals deprived of endogenous glucocorticoid.

A further complication is the question of whether the hormone is inducing PNMT in existing SIF cells or is allowing the survival or generation of PNMT-containing SIF cells which would not normally be present. Pertinent to this issue is the finding that dexamethasone administration at P30 does not elicit detectable PNMT staining. However, if animals have been treated with dexamethasone from days P0–6 (with subsequent loss of PNMT staining after hormone withdrawal), glucocorticoid treatment at P30 will cause the reappearance of PNMT (Black et al. 1980). It may be helpful to further examine this question in cultures of dissociated SIF cells. Preliminary results show that the transmitter produced depends on the hormone concentration: cells grown in 10^{-8} M dexamethasone produce only dopamine, while cultures in 10^{-6} M hormone produce dopamine, norepinephrine, and epinephrine (Doupe and Patterson, unpublished work). It remains to be seen whether the steroid is inducing epinephrine production in all of the cells.

Glucocorticoids also influence the catecholamine enzymes TH and DβH. In adult sympathetic neurons and adrenal chromaffin cells these enzymes are selectively induced by presynaptic activity (see Thoenen 1975; Thoenen et al. 1979). This process, termed transsynaptic induction, occurs with a lag of 24–72 h and is due to the synthesis of new enzyme molecules (Joh et al. 1973). It appears to be mediated by acetylcholine interacting with a nicotinic receptor, but does not require the generation of action potentials (Chalazonitis and Zigmond 1980; Otten and Thoenen 1976b). Glucocorticoids modulate this transsynaptic induction. That is, both in vivo and in organ culture, the hormones have no effect alone but can increase the induction elicited by cold stress or carbamylcholine (Otten and Thoenen 1975, 1976a). In addition, glucocorticoids reduce the time required to achieve the induction (Thoenen and Otten 1978).

Transsynaptic induction also plays a role during development. Blockade of presynaptic activity in the superior cervical ganglion by surgical or pharmacological decentralization at P7 prohibits or limits the normal development rise in TH level (Black et al. 1971, 1974; Black and Geen 1974; Hendry 1973; Thoenen et al. 1972). Glucocorticoids may also modulate this induction. Increasing sympathetic activity by short-term cold stress induced higher TH levels in the adrenal medulla as early as the first few postnatal days. On the other hand, TH inducibility in the superior cervical ganglion did not appear until P15–20 (Otten and Thoenen 1975). The differential time of onset of inducibility in the ganglion and the adrenal correlates with the age at which local corticosteroid levels rise in these structures. Furthermore, pretreatment of P9 animals with dexamethasone makes possible cold stress TH induction in the ganglion at this early age (Otten and Thoenen 1975). Accordingly, very young neurons require transsynaptic stimulation for their normal developmental increase in TH, but are not capable of responding to supranormal sympathetic activity without increased glucocorticoid levels.

Glucocorticoids modulate the induction of TH and DβH by NGF as well. In organ-cultured adult ganglia, dexamethasone does not increase TH activity when administered alone, but dramatically increases the NGF-mediated TH induction, shifting the NGF dose-response curve 30- to 50-fold (Otten and Thoenen 1977). It is unclear at this point whether the hormones also potentiate NGF action during development.

Analysis of the relationship between glucocorticoids and NGF during development is complicated by the absolute dependence of the neonatal neurons on NGF for their survival. This makes it difficult to determine if the steroid plays an NGF-unrelated role. One approach to this problem is the use of the clonal cell line, PC12. This line was derived from an adrenal medullary tumor, but has many characteristics of sympathetic neurons (Greene and Tischler 1976). However, in contrast to primary sympathetic neurons, PC12 cells respond with TH induction to dexamethasone alone but not to NGF alone (Edgar and Thoenen 1978; Goodman et al. 1978). It is further puzzling that at low corticosteroid concentrations ($10^{-8}-10^{-9}$ M), NGF potentiates the steroid effect (Otten and Towbin 1980). Dissociated cell cultures of adrenal chromaffin cells may be very useful for the study of these questions. They are NGF-responsive primary cells but do not require NGF for survival. Unsicker et al. (1978) found that dexamethasone alone did not produce a significant increase in TH activity but slightly potentiated NGF-induced increases in TH.

The site of action of glucocorticoids within the superior cervical ganglion is not yet completely clear, since most biochemical work to date has been done with heterogeneous cell populations in organ culture and in vivo. However, a recent autoradiographic study on the uptake and concentration of ^{3}H-dexamethasone in the adult rat ganglion showed a pronounced and specific labeling only of Schwann and satellite cells (Warembourg et al. 1981). Thus there is as yet no compelling evidence for direct glucocorticoid action on developing sympathetic neurons.

The mechanism of glucocorticoid potentiation of NGF is also unknown, but there is information on the dexamethasone potentiation of another polypeptide hormone, epidermal growth factor (EGF), which stimulates proliferation of fibroblasts. The steroid increases the ability of the cells to bind EGF, perhaps by a change in affinity of the EGF receptors (Baker et al. 1978).

Finally, are catecholamine synthetic enzymes in the brain regulated similarly to those of the periphery? PNMT- and epinephrine-containing nerve tracts have been found in the central nervous system, particularly in the brain stem and hypothalamus (Hökfelt et al. 1973; Saavedra et al. 1974). Neonatal dexamethasone administration increases PNMT activity in both hypothalamus and brain stem, measured both acutely (Moore and Phillipson 1975) or later, in adult animals (Turner et al. 1979). In the latter case, PNMT activity was measured after 30 min of stress, so it is not yet clear if the elevation represents an exaggerated stress response or a permanent elevation in enzyme activity. It is not known if the PNMT increase in brain is due to an enhanced induction of the enzyme as in the adrenal and/or to hyperplasia of inducible PNMT-containing cells as may be occurring in sympathetic ganglia. Recently, Markey et al. (1980) have also reported a 50% increase in TH activity in the pontine region if corticosterone is administered to rats during the 2nd postnatal week, although no response is elicited if steroid is administered later. Thus there is evidence that corticosteroids can affect central catecholamine enzymes during development, but much work remains to be done.

4 Transmitter Choice

A new type of glucocorticoid effect on transmitter development has been reported recently. It is possible to experimentally control the transmitter choice of sympathetic neurons through manipulation of the cellular and hormonal environment surrounding them, both in vivo and in culture (Bunge et al. 1978; LeDouarin 1980; Patterson 1978). Virtually all of the neurons in the neonatal rat superior cervical ganglion display catecholamine fluorescence (Eränkö 1972), and when these cells are dissociated and placed in cell culture, all of the varicosities initially formed contain high proportions of small granular vesicles, indicative of the presence of catecholamine (Johnson et al. 1976; Landis 1980). Similarly, biochemical assays of transmitter production at early times in culture demonstrate the ability of the neuron to synthesize and accumulate catecholamine, but little or no acetylcholine (Patterson and Chun 1977b). This noradrenergic differentiation in vitro is greatly enhanced, or stabilized, by growing the neurons under depolarizing conditions or by stimulating them with action potentials (Landis 1980; Walicke et al. 1977; Walicke and Patterson 1981). On the other hand, these same sympathetic neurons can be influenced to become cholinergic by both diffusible (Landis 1980; Patterson and Chun 1977a) and surface-bound (Hawrot 1980) factors from certain types of non-neuronal cells; the neurons develop the capacity to produce acetylcholine and form functional cholinergic synapses on each other and with cardiac and skeletal muscle cells (Ko et al. 1976; O'Lague et al. 1978; Furshpan et al. 1976; Nurse and O'Lague 1975). Strong cholinergic induction also results in a concomitant lowering of catecholamine synthesis and accumulation (Patterson and Chun 1977a) and endogenous catecholamine content (Landis 1980).

In a search for factors able to prevent the action of such cholinergic signals, McLennan et al. (1980) showed that corticosterone inhibits the development of cholinergic properties in the ganglion in organ culture. Normally after 14 days in culture, choline acetyltransferase activity is high and TH levels have fallen; $10^{-6}\,M$ corticosterone in the medium prevents this change and decreases the choline acetyltransferase/TH activity ratio tenfold. In organ culture it is not possible to tell whether the effect of glucocorticoid is directly on the neurons or on the non-neuronal cells, nor whether the hormone selectively destroys cholinergic neurons. Fukada (1980) examined these questions in dissociated cell culture of sympathetic neurons. By developing a serum-free medium for the preparation of the diffusible cholinergic factor in heart-cell conditioned medium she was able to determine the effect of addition of a variety of hormones on transmitter choice. Conditioned medium made from heart cells incubated with serum-free medium plus hydrocortisone was strikingly ineffective in inducing synthesis of acetylcholine when placed on the neurons. However, addition of hydrocortisone to the neurons along with serum-free conditioned medium (made without hydrocortisone) allowed normal cholinergic induction, demonstrating that the steroid acts on the heart cells to inhibit the production or release of the cholinergic factor. This hormonal effect is not due to a gross deleterious effect on the heart cells: DNA content, total protein, and protein synthesis did not change during the 24 h period of incubation with glucocorticoid for conditioned medium collection. Thus the hormone action is rather specific. Recent work suggests that the sympathetic cholinergic neurons innervating sweat glands in vivo may go through the same adrenergic-to-cholinergic transition seen in culture (Landis and Keefe 1980; Landis 1981). This may prove to be an excellent system in which to study glucocorticoid control of transmitter choice in vivo.

A similar phenomenon may occur in the embryonic gut, where a population of neuron-like cells transiently express noradrenergic properties (TH and DβH immunofluorescence and formaldehyde-induced catecholamine fluorescence) (Cochard et al. 1978; Teitelman et al. 1978). Although it is not known if these cells subsequently die, it is possible that they undergo a transition similar to that described above and adopt a new transmitter phenotype (Jonakait et al. 1979). Therefore it is interesting that hydrocortisone administration to pregnant mothers results in a persistence of the noradrenergic characteristics for a period beyond the normal time of their disappearance (Jonakait et al. 1980). The glucocorticoids may simply be inducing adrenergic enzymes to higher levels, or they could be influencing the transmitter choice. Work by a Russian group (Korochkin and Korochkina 1970) could be interpreted in a similar light. They found that early administration of ACTH or corticosteroid stimulates the development of the sympathetic nervous system, and during the 3rd postnatal week appears to inhibit parasympathetic growth.

These actions of glucocorticoids on adrenergic properties are qualitatively different from those on the TH induction previously considered. That is, as with the neuron-SIF cell-chromaffin cell relationship, the corticosteroids can influence a choice between phenotypes, as opposed to accelerating the expression of one phenotype previously determined. It is also noteworthy that, as will be seen for a number of glucocorticoid effects in the following sections, the hormonal action affecting transmitter choice is on non-neuronal cells, not neurons.

5 Glutamine Synthetase

One of the better understood cases of glucocorticoid effects on central nervous system development is that of the enzyme glutamine synthetase (GS). This enzyme catalzyes the amination of glutamate to glutamine and is important in the pathway for salvaging glutamate (Starr 1974). In the neural retina of the chick embryo, GS activity is low during early development. It increases slowly until E16–17, when it starts to rise sharply, achieving a stable specific activity more than 100-fold above that in the early embryo by E24 (Piddington and Moscona 1965). The onset of the sharp rise in GS activity occurs shortly after the elevation of systemic corticosteroids (Koehler and Moscona 1975). Moreover, retinal GS can be induced precociously by injection of steroids into the embryo or by culture of retinal explants in medium containing 11-β-hydroxycorticosteroid (Piddington 1967). The organ culture studies showed not only that the steroid is acting directly on the retina, but that the enzyme can be induced as early as E8 (Piddington and Moscona 1967). The induced increase in GS activity is due to enzyme synthesis and accumulation: it requires RNA and protein synthesis and involves an accumulation of stable mRNA templates for GS (Moscona et al. 1968). The result is a rapid increase in the rate of de novo synthesis of GS, without direct activation of precursors or changes in the degradation rate (Moscona et al. 1972). The induction process is dependent on the continued presence of the hormone, although GS levels do not fall but simply plateau with steroid withdrawal (Moscona et al. 1972).

The inducibility of GS by glucocorticoids is not constant throughout development: the competence for induction is fully present at E8–9, but at E5–6 the enzyme is essentially not inducible (Moscona and Moscona 1979). However, the embryonic retina can take up corticosteroid and contains abundant cytoplasmic steroid receptor at E6, and the level of receptor does not further increase with time (Koehler and Moscona 1975; Lippman et al. 1974). Thus the control of induction may not be at the level of cytoplasmic receptor concentration.

Recently, GS has been localized by immunohistochemical staining (Linser and Moscona 1979). In the mature retina and in hormone-induced retina and retinal organ cultures, GS was detected solely within Müller fibers, the retinal glial cells. This raises the interesting question of whether these cells also contain the steroid receptors, or whether the induction of GS is secondary to a steroid effect on another cell type. The autoradiographic and culture studies that would answer this question have not yet been performed. However, it has been suggested that GS induction depends on cellular interactions. In low density dissociated cell cultures of trypsinized retina from E8–10 embryos, hydrocortisone does not elicit GS induction, whereas dissociated cells which are allowed to reaggregate immediately after trypsinization do show GS inducibility (Morris and Moscona 1970). It is possible, nonetheless, that the hormone-responsive cells simply die in the low density cultures, such cultures being more subject to toxic components and inadequate nutrients than reaggregate or explant cultures. It will be important to see if reaggregation of cells after growth in low density culture for several days results in reacquisition of GS inducibility.

Possibly relevant in this context are studies on the sensitivity of GS induction to bromodeoxyuridine (BrdU) (Moscona and Moscona 1979). This thymidine analog is incorporated into DNA and irreversibly suppresses the development of GS inducibility by corticosteroids when applied to retinal explants of E5 embryos for 24 h. On the other hand, application of BrdU for 24 h to E8 retinas does not affect the development of GS inducibility. Since BrdU at E5 causes defective histogenesis resulting in drastic malformation of the tissue, it has been suggested that the BrdU-induced loss of GS induction might be secondary to the loss of the normal histotypic organization of the retina. It is difficult, however, to rule out the hypothesis that BrdU simply has deleterious effects on the Müller cells. Evaluation of the role of cellular interactions in the GS response to corticosteroid could be enhanced through studies on Müller cell growth and differentiation in cell cultures using neuronal and glial markers (Barnstable 1980; Raff et al. 1979) and the anti-GS antiserum.

6 Glycerol Phosphate Dehydrogenase

The cytoplasmic enzyme, glycerol phosphate dehydrogenase (GPDH), is thought to be involved in phospholipid synthesis and/or intracellular hydrogen ion transport (see de Vellis and Kukes 1973) and is found in a variety of tissues, including brain, liver, and skeletal muscle. Only in brain, however, does this enzyme appear to be regulated by corticosteroids. In adult rats, hypophysectomy or adrenalectomy causes a 60% decrease in GPDH activity, while total brain DNA and protein contents are unchanged (de Vellis and Inglish 1968). Administration of ACTH or glucocorticoid prevents this decrease in enzyme activity, although these agents do not directly activate the enzyme molecule itself.

During normal postnatal development, brain GPDH activity increases five- to seven-fold, most of the increase occurring between P15 and P35, which coincides with the period of active myelination (de Vellis et al. 1967). Hypophysectomy of P20 rats prevents the normal developmental increase in brain GPDH activity, without a change in protein content (de Vellis and Inglish 1968). Hydrocortisone injections during the 1st or 2nd postnatal weeks cause increased GPDH activity in cerebellum and brain-stem, although not in cerebrum (de Vellis and Inglish 1973). This increased activity does not exceed the normal adult level, suggesting that exogenous glucocorticoid induces the precocious appearance of adult levels of GPDH activity at a time when the rat's own pituitary-adrenal axis is still nonfunctional and blood corticosteroid levels are low.

Induction of GPDH activity is a direct effect of the adrenocortical hormone on brain, as demonstrated in explant cultures: with hormone addition GPDH levels are increased in explants from a variety of E21 rat brain regions, including the cerebrum, after as little as 4 days in vitro. Inducibility increases with time in culture to a maximum at about 15–21 days in vitro (Breen and de Vellis 1975). The induction is inhibited by cycloheximide and actinomycin D, but not by DNA synthesis blockers.

Similar increases in GPDH activity with hydrocortisone have been documented in dissociated cultures of fetal rat cerebrum (Breen and de Vellis 1974). Although these cultures are heterogeneous, containing neurons and non-neuronal cells including glia, the age of maximum inducibility coincides with the age at which the cultures have essentially no veratridine-blocked (presumably neuronal) sodium uptake activity, suggesting that the enzyme is glial (de Vellis et al. 1977). Withdrawal of the hormone from the developing cultures results in a return of GPDH to basal levels. Thus both the maintained adult levels of GPDH and its developmental increase are dependent on the continuous presence of glucocorticoid, as is the case for PNMT but not GS.

The GPDH induction has also been shown in the rat glioma cell line, C6 (McGinnis and de Vellis 1974), another suggestion that the enzyme is glial. Most of the further characterization of the glucocorticoid effect has been done with these cells. It has been demonstrated that the increase in GPDH activity is due to an increase in the number of enzyme molecules. Furthermore, the new molecules are of the same isozymic form as the basal level enzyme (McGinnis and de Vellis 1974). The increase in enzyme is the result of a higher rate of synthesis and not a slower rate of degradation (McGinnis and de Vellis 1976). A high-affinity corticosteroid receptor binding activity has been detected in the cytosol of the C6 cells, and a preliminary report has correlated GPDH inducibility with the concentration of this receptor protein and with the amount of the hormone-receptor complex in the cell nucleus (de Vellis et al. 1974). Only glucocorticoids produce the induction, and progesterone, which competes with corticosteroids for receptor binding, inhibits the GPDH induction (de Vellis et al. 1971). Recently, hydrocortisone induction of GPDH activity has been shown in reportedly pure cultures of primary oligodendrocytes (Weingarten and de Vellis 1980). Thus it is likely that these cells have both corticosteroid receptors and GPDH.

Studies on the localization of the enzyme in vivo have begun, using an antiserum to GPDH. In a preliminary report, electron and light microscope immunhistochemistry has been used to demonstrate that the GPDH-positive cells are, by morphological criteria, oligodendroglia (Leveille et al. 1977). These localization results, taken together with the coincidence in time of the normal increase in GPDH and the period of myelination, and with the suggested involvement of GPDH in lipid synthesis, raise the possibility that this enzyme and the glucocorticoids that control it are important in myelination. Hydrocortisone also increases the activity in C6 glioma cells of $2',3'$-cyclic nucleotide $3'$-phosphohydrolase (Waziri and Sahu 1980), another enzyme thought to be involved in myelin synthesis (Matthieu et al. 1978).

On the other hand, glucocorticoid administration to developing rats inhibits myelination (although this may not be a direct hormonal effect) (Friedrich and Bohn 1980). These observations suggest that if glucocorticoid hormone is involved in the regulation of myelination, its concentration and the timing of its increases must be important. With recent advances in the techniques for identifying and culturing Schwann cells and oligodendrocytes capable of producing myelin-specific proteins (Mirsky et al. 1980), excellent systems are now available for the study of such questions.

7 Other Enzymes

PNMT is one of a number of enzymes in the catecholamine and indoleamine pathways which are methyltransferases and utilize S-adenosyl methionine as a methyl donor. Another of these enzymes is hydroxyindole-O-methyltransferase (HIOMT), which is found in the pineal gland and is responsible for melatonin synthesis. Evidence now suggests that it may be regulated by glucocorticoids in a manner similar to PNMT. Hypophysectomy decreases HIOMT levels in the adult rat pineal, and both dexamethasone and S-adenosyl methionine administration restore enzyme levels to control values, although they have no effect in intact rats (Sandrock et al. 1980). It remains to be established whether HIOMT degradation is accelerated by the absence of glucocorticoids, but the initial similarities between the control of PNMT and HIOMT in adults are striking. During development, HIOMT is first detected at E10 in the chick pineal gland. Its activity increases rapidly just before hatching, concurrently with increases of plasma corticosteroids and thyroid hormone (Wainwright 1974). In organ culture studies of embryonic pineal glands, mixtures of the hormones hydrocortisone, somatotropin, and thyroxine stimulated increased HIOMT activity, although the effect of hydrocortisone alone was not marked. As with PNMT in the developing medulla, some increase in enzyme activity occurred in the absence of any of the hormones (Mezei and Wainwright 1979).

Glucocorticoids may also be involved in the development of brain tryptophan hydroxylase (TrpH), the rate-limiting enzyme in serotonin biosynthesis. Some studies have reported that adult TrpH activity is reduced by adrenalectomy (Azmitia and

McEwen 1969), although this is controversial (Neckers and Sze 1975). It is clear, however, that, as with GPDH, TrpH activity is low in newborn rats and increases during development to reach a steady level by 30 days (Schmidt and Sanders-Bush 1971). Bilateral adrenalectomy in Pg rats completely prevents the normal developmental increase. Glucocorticoid replacement restores the activity and actually induces it slightly, while another tryptophan-related enzyme, 5-hydroxytryptophan decarboxylase, is unaffected (Sze et al. 1976). The normal developmental increase of TrpH correlates well with the normal rise in brain corticosterone levels. As with GS, there is an early time, P1–5, when inducibility is not seen (Sze 1976). In normal adult rats, administration of large doses of corticosteroid does not affect the enzyme level, but the presence of the hormone is required for the action of a variety of other agents which increase TrpH levels (ethanol intoxication, reserpine, footshock) (Sze 1976). At present there is no evidence that this increase in enzyme activity requires protein synthesis, nor that the steroid is acting directly on neurons, but the developmental history of the enzyme is very reminiscent of other glucocorticoid-dependent enzymes described above.

8 Conclusions

The cellular and molecular analysis of direct glucocorticoid actions on the developing nervous system is at a very early stage. Even so, it is clear that these hormones have widespread effects. That is, they can act on various areas of the brain, the autonomic nervous system, and the adrenal medulla, as well as on a variety of cell types. Autoradiographic studies in the brain suggest a predominantly neuronal localization for corticosteroid receptors, but there is also clear evidence of direct hormone action on glia and chromaffin cells.

It will be very interesting to determine if the neuronal glucocorticoid receptors in the brain, whose function in the adult is unknown, play a developmental role. In addition to having a diversity of targets, corticosteroids can act in several different ways: they can (1) play a role in the choice between several different phenotypes (neuron vs chromaffin cell, adrenergic vs cholinergic), (2) accelerate the differentiation and maturation of a previously determined phenotype (PNMT, GPDH, and GS), and (3) modulate the influence of other developmental signals (NGF, presynaptic activity). Each of these types of hormonal control can persist after development is completed, and in some cases the hormone is required throughout adult life for maintenance of full expression of particular characteristics (PNMT, GPDH). Finally, in light of the extensive clinical use of glucocorticoids, it should also be noted that the premature acceleration of development caused by administration of these hormones may have adverse effects.

The work from the authors' laboratory was supported by the NINCDS, the Rita Allen Foundation, the Insurance Medical Scientist Scholarship Fund, and the Prudential Life Insurance Company.

References

Allen C, Kendall JW (1967) Maturation of the circadian rhythm of plasma corticosterone in the rat. Endocrinology 80:926–930

Aloe L, Levi-Montalcini R (1979) Nerve growth factor-induced transformation of immature chromaffin cells in vivo into sympathetic neurons: effect of antiserum to nerve growth factor. Proc Natl Acad Sci USA 76:1246–1250

Azmitia EC, McEwen BS (1969) Corticosterone regulation of tryptophan hydroxylase in the midbrain of the rat. Science 166:1274–1276

Baker JB, Barsh GS, Carney DH, Cunningham DD (1978) Dexamethasone modulates binding and action of epidermal growth factor in serum-free cell culture. Proc Natl Acad Sci USA 75:1882–1886

Ballard PL (1979) Glucocorticoids and differentiation. In: Baxter JD, Rousseau GG (eds) Glucocorticoid hormone action. Springer, Berlin Heidelberg New York, pp 493–515

Barnstable CJ (1980) Monoclonal antibodies which recognize different cell types in the rat retina. Nature 286:231–235

Björklund A, Cegrell L, Falck B, Ritzén M, Rosengren E (1970) Dopamine-containing cells in sympathetic ganglia. Acta physiol Scand 78:334–338

Black IB, Geen SC (1974) Inhibition of the biochemical and morphological maturation of adrenergic neurons by nicotinic receptor blockade. J Neurochem 22:301–306

Black IB, Hendry IA, Iversen LL (1971) Trans-synaptic regulation of growth and development of adrenergic neurones in a mouse sympathetic ganglion. Brain Res 34:229–240

Black IB, Joh TH, Reis DJ (1974) Accumulation of tyrosine hydroxylase molecules during growth and development of the superior cervical ganglion. Brain Res 75: 133–144

Black IB, Bohn MC, Bloom EM, Goldstein M (1980) Glucocorticoids induce expression of the adrenergic phenotype in a rat sympathetic ganglion. Soc Neurosci Abstracts 6:408 (#138.1)

Bohn MC, Goldstein M, Black IB (1980) The role of glucocorticoid steroids in the expression of the adrenergic phenotype in the rat embryonic adrenal gland. Soc Neurosci Abstr 6:644 (#218.5)

Bohn MC, Goldstein M, Black IB (1981) Role of glucocorticoids in expression of the adrenergic phenotype in rat embryonic adrenal gland. Dev Biol 82:1–10

Breen GAM, de Vellis J (1974) Regulation of glycerol phosphate dehydrogenase by hydrocortisone in dissociated rat cerebral cell cultures. Dev Biol 41:255–266

Breen GAM, de Vellis J (1975) Regulation of glycerol phosphate dehydrogenase by hydrocortisone in rat brain explants. Exp Cell Res 91:159–169

Brodsky M, Teitelman G, Park DH, New M, Joh TH, Reis DJ (1980) The expression of an adrenergic phenotype in fetal adrenal medullary cells is not induced by glucocorticoids. Soc Neurosci Abstr 6:409 (#138.6)

Bunge RP, Johnson M, Ross CD (1978) Nature and nurture in development of the autonomic neuron. Science 199:1409–1416

Burnstock G, Costa M (1975) Adrenergic neurons. Chapman & Hall, London

Chalazonitis A, Zigmond RE (1980) Effects of synaptic and antidromic stimulation on tyrosine hydroxylase activity in the rat superior cervical ganglion. J Physiol (Lond) 300:525–538

Ciaranello RD (1978) Regulation of phenylethanolamine-N-methyltransferase synthesis and degradation. I. Regulation by rat adrenal glucocorticoids. Mol Pharmacol 14:478–489

Ciaranello RD, Jacobowitz D, Axelrod J (1973) Effect of dexamethasone on phenylethanolamine-N-methyltransferase in chromaffin tissue of the neonatal rat. J Neurochem 20:799–805

Ciaranello RD, Wong DL, Berenbeim DM (1978) Regulation of phenylethanolamine-N-methyltransferase synthesis and degradation. II. Control of the thermal stability of the enzyme by an endogenous stabilizing factor. Mol Pharmacol 14:490–501

Cochard P, Goldstein M, Black IB (1978) Ontogenetic appearance and disappearance of tyrosine hydroxylase and catecholamines in the rat embryo. Proc Natl Acad Sci USA 75:2986–2990

Cohen AM (1972) Factors directing the expression of sympathetic nerve traits in cells of neural crest origin. J Exp Zool 179:167–182

Cotterrell M, Balázs R, Johnson AL (1972) Effect of corticosteroids on the biochemical maturation of rat brain: postnatal cell formation. J Neurochem 19:2151–2167

Coupland RE (1965a) The natural history of the chromaffin cell. Longmans & Green, London

Coupland RE (1965b) Electron microscopic observations on the structure of the rat adrenal medulla. I. The ultrastructure and organization of chromaffin cells in the normal adrenal medulla. J Anat 99:231–254

Coupland RE (1975) Blood supply of the adrenal gland. In: Blaschko H, Sayes G, Smith AD (eds) Adrenal gland. Amer Physiol Soc, Washington (Handbook of physiology, vol VI, sect 7, pp 283–294)

Coupland RE, MacDougall JDB (1966) Adrenalin formation in noradrenaline-storing chromaffin cells in vitro induced by corticosterone. J Endocrinology 36:317–324

de Vellis J, Inglish D (1968) Hormonal control of glycerol phosphate dehydrogenase in the rat brain. J Neurochem 15:1061–1070

de Vellis J, Inglish D (1973) Age-dependent changes in the regulation of glycerol phosphate dehydrogenase in the rat brain and in a glial cell line. Prog Brain Res 40:321–330

de Vellis J, Kukes G (1973) Regulation of glial cell functions by hormones and ions: a review. Tex Rep Biol Med 31:271–293

de Vellis J, Schjeide OA, Clemente CD (1967) Protein synthesis and enzymic patterns in the developing brain following head X-irradiation of newborn rats. J Neurochem 14:499–511

de Vellis J, Inglish D, Cole R, Molson J (1971) Effects of hormones on the differentiation of cloned lines of neurons and glial cells. In: Ford E (ed) Influence of hormones on the nervous system. Karger, Basel, pp 25–39

de Vellis J, McEwen BS, Cole R, Inglish D (1974) Relations between glucocorticoid nuclear binding, cytosol receptor activity and enzyme induction in a rat glial cell line. J Steroid Biochem 5:392–393

de Vellis J, McGinnis JF, Breen GAM, Leveille P, Bennett K, McCarthy K (1977) Hormonal effects on differentiation in neural cultures. In: Federoff S, Hertz L (eds) Cell, tissue and organ culture in neurobiology. Academic Press, New York, pp 485–511

Doupe AJ, Patterson PH, Landis SC (1980) Dissociated cell culture of SIF cells: hormone-dependence and NGF action. Soc Neurosci Abstr. 6:409 (#138.5)

Edgar D, Thoenen H (1978) Selective enzyme induction in a nerve growth factor-responsive pheochromocytoma cell line. Brain Res 154:186–190

Elfvin LG, Hökfelt T, Goldstein M (1975) Fluorescence microscopical, immunohistochemical and ultrastructural studies on sympathetic ganglia of the guinea pig, with special reference to the SIF cells and their catecholamine content. J Ultrastruc Res 51:377–396

Eränkö L (1972) Postnatal development of histochemically demonstrable catecholamines in the superior cervical ganglion of the rat. Histochem J 4:225–236

Eränkö L, Eränkö O (1972) Effect of hydrocortisone on histochemically demonstrable catecholamines in the sympathetic ganglia and extra-adrenal chromaffin tissue of the rat. Acta Physiol Scand 84:125–133

Eränkö O, Lempinen M, Raisanen L (1966) Adrenaline and noradrenaline in the organ of Zuckerkandl and adrenals of newborn rats treated with hydrocortisone. Acta Physiol Scand 66:253–254

Eränkö O, Eränkö L, Hill CE, Burnstock G (1972a) Hydrocortisone-induced increase in the number of small intensely fluorescent cells and their histochemically demonstrable catecholamine content in cultures of sympathetic ganglia of the newborn rat. Histochem J 4:49–58

Eränkö O, Heath J, Eränkö L (1972b) Effect of hydrocortisone on the ultrastructure of the small, intensely fluorescent, granule-containing cells in cultures of sympathetic ganglia of newborn rat. Z Zellforsch 134:297–310

Friedrich VL, Bohn MC (1980) Glucocorticoids inhibit myelination in developing rat. Soc Neurosci Abstr 6:380 (#132.3)

Fukada K (1980) Hormonal control of neurotransmitter choice in sympathetic neurone cultures. Nature 287:553–555

Furness JB, Costa M (1976) Some observations on extra-adrenal chromaffin cells of the lower abdomen and pelvis. In: Coupland RE, Fujita T (eds) Chromaffin, enterochromaffin and related cells. Elsevier, Amsterdam, pp 25–34

Furshpan EJ, MacLeish PR, O'Lague PH, Potter DD (1976) Chemical transmission between rat sympathetic neurons and cardiac myocytes developing in microcultures: evidence for cholinergic, adrenergic and dual-function neurons. Proc Natl Acad Sci USA 73:4225–4229

Fuxe K, Goldstein M, Hökfelt T, Joh TH (1971) Cellular localization of dopamine-β-hydroxylase and phenylethanolamine-N-methyltransferase as revealed by immunohistochemistry. Prog Brain Res 34:127–138

Goodman R, Edgar D, Thoenen H, Wechsler W, Herschman H (1978) Glucocorticoid induction of tyrosine hydroxylase in a continuous cell line of rat pheochromocytoma. J Cell Biol 78:R1–R7

Greene LA, Tischler AS (1976) Establishment of a noradrenergic clonal line of rat adrenal pheochromocytoma cells which respond to nerve growth factor. Proc Natl Acad Sci USA 73:2424–2428

Gundersen RW, Barrett JN (1979) Neuronal chemotaxis: chick dorsal root axons turn toward high concentrations of nerve growth factor. Science 206:1079–1080

Hawrot E (1980) Cultured sympathetic neurons: effects of cell-derived and synthetic substrata on survival and development. Dev Biol 74:136–151

Hendry IA (1973) Trans-synaptic regulation of tyrosine hydroxylase activity in a developing mouse sympathetic ganglion: effects of nerve growth factor (NGF), NGF-antiserum, and pempidine. Brain Res 56:313–320

Hökfelt T, Fuxe K, Goldstein M, Johansson O (1973) Evidence for adrenaline neurons in the rat brain. Acta Physiol Scand 89:286–289

Howard E (1968) Reductions in size and total DNA of cerebrum and cerebellum in adult mice after corticosterone treatment in infancy. Exp Neurol 22:191–208

Jacobson M (1978) Developmental neurobiology. Plenum Press, New York, pp 219–224

Joh TH, Geghman C, Reis DJ (1973) Immunochemical demonstration of increased accumulation of tyrosine hydroxylase protein in sympathetic ganglia and adrenal medulla elicited by reserpine. Proc Natl Acad Sci USA 70:2767–2771

Johnson LK, Baxter JD, Rousseau GG (1979) Mechanisms of glucocorticoid receptor function. In: Baxter JD, Rousseau GG (eds) Glucocorticoid hormone action. Springer, Berlin Heidelberg New York, pp 305–326

Johnson M, Ross D, Meyers M, Rees R, Bunge R, Wakshull E, Burton H (1976) Synaptic vesicle cytochemistry changes when cultured sympathetic neurons develop cholinergic interactions. Nature 262:308–310

Jonakait GM, Wolf J, Cochard P, Goldstein M, Black IB (1979) Selective loss of noradrenergic phenotypic characters in neuroblasts of the rat embryo. Proc Natl Acad Sci USA 76:4683–4686

Jonakait GM, Bohn MC, Black IB (1980) Maternal glucocorticoid hormones influence neurotransmitter phenotypic expression in embryos. Science 210:551–553

Jones MT, Hillhouse EW, Burden JL (1977) Dynamics and mechanics of corticosteroid feedback at the hypothalamus and anterior pituitary gland. J Endocrinol 73:405–417

Jost A (1966) Problems of fetal endocrinology: the adrenal glands. Recent Prog Horm Res 22:541–574

Ko C-P, Burton H, Johnson MI, Bunge RP (1976) Synaptic transmission between rat superior cervical ganglion neurons in dissociated cell cultures. Brain Res 117: 461–485

Koehler DE, Moscona AA (1975) Corticosteroid receptors in the neural retina and other tissues of the chick embryo. Arch Biochem Biophys 170:102–113

Korochkin LI, Korochkina LS (1970) Hormonal influence on the differentiation of nerve cells of sympathetic and parasympathetic nervous system. Z Mikrosk Anat Forsch 82:293–321

Koslow SH, Bjegovic M, Costa E (1975) Catecholamines in sympathetic ganglia of rat: Effects of dexamethasone and reserpine. J Neurochem 24:277–281

Landis SC (1980) Developmental changes in the neurotransmitter properties of dissociated sympathetic neurons: a cytochemical study of the effects of medium. Dev Biol 77:349–361

Landis SC (1981) Environmental influences on the postnatal development of rat sympathetic neurons. In: Garrod DR, Feldman JD (eds) Development in the nervous system. Cambridge University Press, Cambridge, pp 147–160

Landis SC, Keefe D (1980) Development of cholinergic sympathetic innervation of ecrine sweat glands in rat footpad. Soc Neurosci Abstr 6:379 (#131.20)

LeDouarin NM (1980) The ontogeny of the neural crest in avian embryo chimeras. Nature 286:663–669

Lempinen M (1964) Extra-adrenal chromaffin tissue of the rat and the effect of cortical hormones on it. Acta Physiol Scand (Suppl 231) 62:1–91

Leveille PJ, de Vellis J, Maxwell DS (1977) Immunocytochemical localization of glycerol-3-phosphate dehydrogenase in rat brain: are oligodendrocytes target cells for glucocorticoids? Soc Neurosci Abstr 3:333 (#1066)

Liggins GC (1976) Adrenocortical-related maturational events in the fetus. Am J Obstet Gynecol 126:931–941

Linser P, Moscona AA (1979) Induction of glutamine synthetase in embryonic neural retina: localization in Müller fibres and dependence on cell interactions. Proc Natl Acad Sci USA 76:6476–6480

Lippman ME, Wiggert BO, Chader GJ, Thompson BE (1974) Glucocorticoid receptors. Characteristics, specificity, and ontogenesis in the embryonic chick neural retina. J Biol Chem 249:5916–5917

Lu KS, Lever JD, Santer RM, Presley R (1976) Small granulated cell types in rat superior cervical and coeliac-mesenteric ganglia. Cell Tiss Res 172:331–343

Margolis FL, Roffi J, Jost A (1966) Norepinephrine methylation in fetal rat adrenals. Science 154:275–276

Markey KA, Towle AC, Sze PY (1980) Glucocorticoid effects on brain tyrosine hydoxylase. Soc Neurosci Abstr 6:144 (#53.27)

Martin CE, Cake MH, Hartmann PE, Cook IR (1977) Relationship between fetal corticosteroids, maternal progesterone and parturition in the rat. Acta Endocrinol 84: 167–176

Matthieu J-M, Honegger P, Trapp BD, Cohen SR, Webster HdeF (1978) Myelination in rat brain aggregating cell cultures. Neuroscience 3:565–572

McEwen BS (1978) Influences of adrenocortical hormones on pituitary and brain function. In: Baxter JD, Rousseau GG (eds) Glucocorticoid hormone action. Springer, Berlin Heidelberg New York, pp 467–492

McEwen BS, Gerlach JL, Micco DJ (1975) Putative glucocorticoid receptors in hippocampus and other regions of the rat brain. In: Isaacson R, Pribram K (eds) The hippocampus: a comprehensive treatise. Plenum Press, New York, pp 285–322

McEwen BS, Davis PG, Parsons B, Pfaff DW (1979) The brain as a target for steroid hormone action. Ann Rev Neurosci 2:65–112

McGinnis JF, de Vellis J (1974) Cortisol induction of glycerol phosphate dehydrogenase in a rat brain tumor cell line. Nature 250:422–424

McGinnis JF, de Vellis J (1976) Glucocorticoid regulation of the concentration of glycerolphosphate dehydrogenase in a rat glioma cell line. Fed Proc 35:1636

McLennan IS, Hill CE, Hendry IA (1980) Glucocorticosteroids modulate transmitter choice in developing superior cervical ganglion. Nature 283:206–207

Mezei C, Wainwright SD (1979) Hormone-induced increase of hydroxyindole-O-methyltransferase activity in the embryonic chick pineal gland in organ culture. Life Sci 24:1111–1117

Mirsky R, Winter J, Abney ER, Pruss RM, Gavrilovic J, Raff MC (1980) Myelin-specific proteins and glycolipids in rat Schwann cells and oligodendrocytes in culture. J Cell Biol 84:483–494

Moore KE, Phillipson OT (1975) Effects of dexamethasone on phenylethanolamine-N-methyltransferase and adrenaline in the brains and superior cervical ganglia of adult and neonatal rats. J Neurochem 25:289–294

Morris JE, Moscona AA (1970) Induction of glutamine synthetase in embryonic retina: its dependence on cell interactions. Science 167:1736–1738

Moscona AA, Moscona MH, Saenz N (1968) Enzyme induction in embryonic retina: the role of transcription and translation. Proc Natl Acad Sci USA 61:160–167

Moscona M, Moscona AA (1979) The development of inducibility for glutamine synthetase in embryonic neural retina: inhibition by BrdU. Differentiation 13:165–172

Moscona M, Frenkel N, Moscona AA (1972) Regulatory mechanisms in the induction of glutamine synthetase in the embryonic retina: immunochemical studies. Dev Biol 28:229–241

Mulay S, Giannopoulos G, Solomon S (1973) Corticosteroid levels in the mother and fetus of the rabbit during gestation. Endocrinology 93:1342–1348

Neckers L, Sze PY (1975) Regulation of 5-hydroxytryptamine metabolism in mouse brain by adrenal glucocorticoids. Brain Res 93:123–132

Nurse CA, O'Lague PH (1975) Formation of cholinergic synapses between dissociated sympathetic neurons and skeletal myotubes of the rat in cell culture. Proc Natl Acad Sci USA 72:1955–1959

O'Lague PH, Potter DD, Furshpan EJ (1978) Studies on rat sympathetic neurons developing in cell culture. III. Cholinergic transmission. Dev Biol 67:424–443

Olpe H-R, McEwen BS (1976) Glucocorticoid binding to receptor-like proteins in rat brain and pituitary: ontogenetic and experimentally induced changes. Brain Res 105:121–128

Olson L (1970) Fluorescence histochemical evidence for axonal growth and secretion from transplanted adrenal medullary tissue. Histochemie 22:1–7

Otten U, Thoenen H (1975) Circadian rhythm of tyrosine hydroxylase induction by short-term cold stress: modulatory actions of glucocorticoids in newborn and adult rat. Proc Natl Acad Sci USA 72:1415–1419

Otten U, Thoenen H (1976a) Selective induction of typrosine hydroxylase and dopamine-β-hydroxylase in sympathetic ganglia in organ culture: role of gluco-corticoids as modulators. Mol Pharmacol 12:353–361

Otten U, Thoenen H (1976b) Role of membrane depolarization in transsynaptic induction of tyrosine hydroxylase in organ cultures of sympathetic ganglia. Neurosci Lett 2:93–96

Otten U, Thoenen H (1977) Effect of glucocorticoids on NGF-mediated enzyme induction in organ cultures of rat sympathetic ganglia: enhanced response and reduced time requirement to initiate enzyme induction. J Neurochem 29:69–75

Otten U, Towbin M (1980) Permissive action of glucocorticoids in induction of tyrosine hydroxylase by nerve growth factor in a pheochromocytoma cell line. Brain Res 193:304–308

Patterson PH (1978) Environmental determination of autonomic neurotransmitter functions. Ann Rev Neurosci 1:1−17

Patterson PH, Chun LLY (1977a) The induction of acetylcholine synthesis in primary cultures of dissociated rat sympathetic neurons. I. Effects of conditioned medium. Develop Biol 56:263−280

Patterson PH, Chun LLY (1977b) The induction of acetylcholine synthesis in primary cultures of dissociated rat sympathetic neurons. II. Developmental aspects. Dev Biol 60:473−481

Piddington R (1967) Hormonal effects on the development of glutamine synthetase in the embryonic chick retina. Dev Biol 16:168−188

Piddington R, Moscona AA (1965) Correspondence between glutamine synthetase activity and differentiation in the embryonic retina in situ and in culture. J Cell Biol 27:247−252

Piddington R, Moscona AA (1967) Precocious induction of retinal glutamine synthetase by hydrocortisone in the embryo and in culture: age-dependent differences in tissue response. Biochim Biophys Acta 141:429−432

Pohorecky LA, Wurtman RJ (1971) Adrenocortical control of epinephrine synthesis. Pharmacol Rev 23:1−35

Raff MC, Fields KL, Hakomori S-I, Mirsky R, Pruss RM, Winter J (1979) Cell-type specific markers for distinguishing and studying neurons and the major classes of glial cells in culture. Brain Res 174:283−308

Ramaley JA (1974) The changes in basal corticosterone secretion in rats blinded at birth. Experientia 30:827

Rybarczyk KE, Baker HA, Burke JP, Hartman BK, Van Orden LS (1976) Histochemical and immunocytochemical identification of catecholamines, dopamine-β-hydroxylase and phenylethanolamine-N-methyltransferase. In: Eränkö O (ed) SIF cells. United States Government Printing Office, Washington, pp 68−81

Saavedra JM, Palkovits M, Brownstein MJ, Axelrod J (1974) Localisation of phenylethanolamine N-methyltransferase in the rat brain nuclei. Nature 248:695−696

Sandrock AW, Leblanc GG, Wong DL, Ciaranello RD (1980) Regulation of rat pineal hydroxyindole-O-methyltransferase: evidence of S-adenosylmethionine-mediated glucocorticoid control. J Neurochem 35:536−543

Schmidt MJ, Sanders-Bush E (1971) Tryptophan hydroxylase activity in developing rat brain. J Neurochem 18:2549−2551

Shepherd DM, West GB (1951) Noradrenaline and the suprarenal medulla. Br J Pharmacol Chemother 6:665−674

Starr MS (1974) Evidence for the compartmentalization of glutamate metabolism in isolated rat retina. J Neurochem 23:337−344

Sze PY (1976) Glucocorticoid regulation of the serotonergic system of the brain. Adv Biochem Psychopharmacol 15:251−265

Sze PY, Neckers L, Towle AC (1976) Glucocorticoids as a regulatory factor for brain tryptophan hydroxylase. J Neurochem 26:169−173

Taxi J (1979) The chromaffin and chromaffin-like cells in the autonomic nervous system. Int Rev Cytol 57:283−343

Teillet MA, Cochard P, LeDouarin NM (1978) Relative roles of the mesenchymal tissues and of the complex neural tube-notochord on the expression of adrenergic metabolism in neural crest cells. Zoon 6:115−122

Teitelman G, Joh TH, Reis DJ (1978) Transient expression of a noradrenergic phenotype in cells of the rat embryonic gut. Brain Res 158:229−234

Teitelman G, Baker H, Joh TH, Reis DJ (1979) Appearance of catecholamine-synthesizing enzymes during development of rat sympathetic nervous system: possible role of tissue environment. Proc Natl Acad Sci USA 76:509−513

Thoenen H (1975) Trans-synaptic regulation of neuronal enzyme synthesis. In: Iversen LL, Iversen SD, Snyder SH (eds) Handbook of Psychopharmacol, 3rd edn, vol 3. Plenum Press, New York, pp 443−475

Thoenen H, Otten U (1978) Role of adrenocortical hormones in the modulation of synthesis and degradation of enzymes involved in the formation of catecholamines. In: Ganong WF, Martini L (eds) Frontiers in Neuroendocrinology, vol 5. Raven Press, New York, pp 163–184

Thoenen H, Saner A, Kettler R, Angeletti PU (1972) Nerve growth factor and preganglionic cholinergic nerves: their relative importance to the development of the terminal adrenergic neurons. Brain Res 44:593–602

Thoenen H, Otten U, Schwab M (1979) Orthograde and retrograde signals for the regulation of neuronal gene expression: the peripheral sympathetic nervous system as a model. In: Schmidt FO, Worden FG (eds) The neurosciences fourth study program. MIT Press, Cambridge, pp 911–928

Turner BB, Katz RJ, Carroll BJ (1979) Neonatal corticosteroid permanently alters brain activity of epinephrine-synthesizing enzyme in stressed rats. Brain Res 166: 426–430

Unsicker K, Krisch B, Otten U, Thoenen H (1978) Nerve growth factor-induced fiber outgrowth from isolated rat adrenal chromaffin cells: impairment by glucocorticoids. Proc Natl Acad Sci USA 75:3498–3502

Verhofstad AAJ, Hökfelt T, Goldstein M, Steinbusch HWM, Joosten HWJ (1979) Appearance of tyrosine hydroxylase, aromatic amino-acid decarboxylase, dopamine β-hydroxylase and phenylethanolamine-N-methyltransferase during the ontogenesis of the adrenal medulla. An immunohistochemical study in the rat. Cell Tissue Res 200:1–13

Wainwright SD (1974) Course of the increase in hydroxyindole-O-methyltransferase activity in the pineal gland of the chick embryo and young chick. J Neurochem 22:193–196

Walicke PA, Patterson PH (1981) On the role of Ca^{++} in the transmitter choice made by cultured sympathetic neurons. J Neurosci 1:343–350

Walicke PA, Campenot RB, Patterson PH (1977) Determination of transmitter function by neuronal activity. Proc Natl Acad Sci USA 74:5767–5771

Warembourg M (1975a) Radioautographic study of the rat brain after injection of $[1,2^{-3}H]$ corticosterone. Brain Res 89:61–70

Warembourg M (1975b) Radioautographic study of the rat brain and pituitary after injection of ^{3}H-dexamethasone. Cell Tissue Res 161:183–191

Warembourg M, Otten U, Schwab ME (1981) Labelling of Schwann and satellite cells by ^{3}H-dexamethasone in a rat sympathetic ganglion and sciatic nerve. Neuroscience 6:1139–1144

Waziri R, Sahu SK (1980) Induction of 2′, 3′-cyclic nucleotide 3′-phosphohydrolase and morphological alterations in C6 glioma cells by dexamethasone, (3-butoxy-4-methoxybenzyl)-2-imidazolinone and prostaglandin E1. In Vitro 16:97–102

Weingarten D, de Vellis J (1980) Selective inhibition by sodium butyrate of the glucocorticoid induction of glycerol phosphate dehydrogenase in glial cultures. Biochem Biophys Res Commun 93:1297–1304

Wurtman RJ, Axelrod J (1965) Adrenaline synthesis: control by the pituitary gland and adrenal glucocorticoids. Science 150:1464–1465

Feedback Actions of Adrenal Steroid Hormones

M.T. JONES, B. GILLHAM[1], B.D. GREENSTEIN[2], U. BECKFORD[1] and
M.C. HOLMES

Contents

1 Introduction

The earliest work on the ways in which corticosteroids might influence the functional
activity of the pituitary adrenocortical system was reported by Ingle and Kendall (1937).
They showed that the administration of adrenocortical extracts to intact rats resulted

Departments of Physiology, Biochemistry[1], and Pharmacology[2], St. Thomas's Hospital
Medical School, Lambeth Palace Road, London SE1 7EH, Great Britain

in adrenocortical atrophy, whilst the simultaneous injection of an anterior pituitary extract with that from the adrenal cortex prevented the atrophic response (Ingle and Kendall 1937; Ingle et al. 1938). Subsequently, Deane and Greep (1946) showed that hypophysectomy results in atrophy of the zonae fasiculata and reticularis but the zona glomerulosa was virtually unaffected. A similar histologic picture was described by Winter et al. (1950) following prolonged treatment with cortisone. Their results suggested that corticosteroid treatment inhibits adrenocorticotrophin (ACTH) secretion and the possibility that ACTH secretion could be regulated by adrenocortical secretions.

In 1948, Sayers and Sayers showed that endogenously released or exogenously administered ACTH causes adrenal ascorbate depletion. They found that the fall in adrenal ascorbate induced by the stress of cold or histamine could be prevented by pretreatment with small doses of corticosteroids. They also demonstrated the quantitative relationship between the severity of the stress and the dosage of steroid, given some hours previously, which was required to inhibit the corticotrophic response. Moreover, in one experiment they showed that the ascorbate depletion induced by the injection of adrenaline was inhibited when the amine was preceded by a 3-min infusion of cortisol. Thus, out of this germane work came some fundamental findings, namely that corticosteroids can exercise both immediate and delayed inhibitory effects on the corticotrophic response, and that the dose of steroid required to prevent the corticotrophic response to stress is directly proportional to the intensity of the stress when the steroid is given hours before the provocative test.

The experiments of Sayers and Sayers demonstrated the probable existence of a corticosteroid negative feedback effect but they did not show that the inhibition derived from physiologically relevant mechanisms. Probably the first clear demonstration of the physiologic nature of the feedback mechanism came from the work of James and co-authors (James et al. 1968). They showed that in man the infusion of cortisol at a rate of 1 mg/h^{-1} (equal to the upper limit of endogenous cortisol secretion rate) caused no net change in the urinary excretion of 17-hydroxycorticosteroids, derived metabolically from cortisol, when this was compared with the excretion on days when saline was administered. This clearly indicated the existence of a feedback mechanism in man that has a speed and sensitivity sufficient to compensate for the changes in plasma cortisol levels induced by administered corticosteroid.

There is now abundant evidence for the existence of a negative feedback mechanism as is shown by the following observations:

1. Adrenalectomy results in an elevation in the concentration of ACTH in the blood of rats, which is greater than that produced by a sham operation, and which is reduced towards the sham operation level by replacement therapy (Dallman et al. 1972; Dallman and Jones 1973). Similarly, patients suffering from adrenocortical atrophy (Addison's

---

The following abbreviations appear in the text:

Ach: acetylcholine; ACTH: adrenocorticotrophin; CRF: corticotrophin-releasing factor; CSF: cerebrospinal fluid; DFB: delayed feedback; DOC: 11-deoxycorticosterone; GABA: gamma amino butyric acid; ^{3}H-DHT: labeled androgen; HE: hypothalamic extract; HPA: hypothalamo-pituitary-adrenocortical; K_d: constant of dissociation; S: 11-deoxycortisol

disease) have raised levels of plasma ACTH which are reduced toward the normal level with replacement therapy (Bethune et al. 1957).

2. The release of ACTH in response to stress is greatly exaggerated in adrenalectomised animals or patients (Dallman et al. 1972; Daly et al. 1974).

3. Pretreatment of animals with physiologic doses of corticosteroids serves to modulate the release of ACTH elicited in response to moderate stress (Dallman and Yates 1969; Jones et al. 1972).

2 Theories on the Nature of Corticosteroid Negative Feedback Mechanisms

2.1 The Variable Set-Point Hypothesis

Yates et al. (1961) found that intravenous injection of small doses of corticosterone in the rat partially or completely inhibited the corticotrophic response to stress applied 15 s after steroid administration. They proposed that stress results in a 're-set of the negative feedback control of plasma corticosteroid concentration to regulate at higher levels'. The theory proposed that in response to stress, ACTH secretion continues until plasma corticosteroid concentration reaches a new threshold. Although this hypothesis overcame many objections to the hypothesis of Sayers, it was disproved when it was shown that plasma corticosteroid levels equal to (or greater than) those found during stress did not inhibit the corticotrophic response to stress (Hodges and Jones 1963; Smelik 1963).

2.2 Fast (Rate-Sensitive) and Delayed (Proportional) Feedback Control

Dallman and Yates (1969), using a continuous infusion of corticosterone, showed that there is a short-delay, rate-sensitive feedback control element associated with the rate of rise of the plasma corticosteroid concentration. An 80% inhibition of the stress response occurred during the phase when the corticosterone concentration was rising at its fastest rate (first 5 min) and this inhibition disappeared during the slower phase (5–20 min). No further inhibition of the stress response was seen until 120 min from the start, when an 85% inhibition became evident. Dallman and Yates suggested that the early inhibition was due to the proposed rate-sensitive (or fast) feedback, and the later inhibition was a proportional (or delayed) feedback mechanism.

It appears, therefore, that ACTH secretion is regulated by two temporally distinct feedback mechanisms: fast feedback, which occurs within minutes of administration at a time of rising plasma corticosteroids, and delayed feedback which occurs later when the plasma corticosteroids may be high, declining or low. An explanation of the pharmacologic mechanism responsible for the 'silent period' between the two phases of corticosteroid negative feedback, based on the Paton rate theory, was proposed (Jones et al. 1972) and computer simulation studies confirmed the likelihood of this explanation (Kaneko and Hiroshige 1976). Furthermore, this implies that the failure of others to demonstrate feedback was probably due to the application of stress during the 'silent period'.

The rate of increase of plasma glucocorticoids required to reduce the corticotrophic response to surgical stress has been defined by Jones et al. (1972), who showed that infusion of corticosterone to give a rate of rise of the steroid in the plasma of 1.3 µg/100 ml per min or greater prevented the response to such stress (Fig. 1). This rate-sensitive feedback has now been demonstrated in several laboratories, but the rate of increase in plasma corticosterone required to reduce the stress response is variable (Abe and Critchlow 1977; Kenko and Hiroshige 1978a).

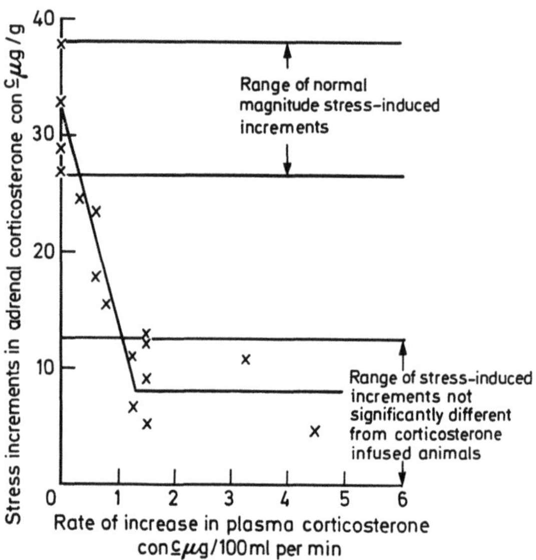

Fig. 1. Relationship between the magnitude of the stress response and the rate of increase of plasma corticosterone. (Jones et al. 1972)

The two principle glucocorticoids secreted by the adrenal gland, cortisol and corticosterone, are fast-feedback agonists and inhibit ACTH secretion during the rising phase in plasma concentration (Jones et al. 1972). The other steroids secreted by the adrenal gland are not effective in this way, which is also true for many synthetic glucocorticoids (Jones et al. 1976; Jones and Tiptaft 1977).

Fast and delayed feedback have different dynamics, and the question arises whether the two mechanisms involve the same or different receptors; this was investigated by testing the effect of different steroids on stress-induced release of ACTH minutes or hours after their administration (Jones et al. 1974; Jones and Tiptaft 1977). The steroids 11-deoxycorticosterone (DOC) and 11-deoxycortisol (S) have no fast-feedback action and indeed antagonize the fast-feedback action of endogenously secreted or exogenously administered corticosterone or cortisol. DOC and S suppress the corticotrophic response to stress if they are given some 2–4 h prior to the stimulus (Jones et al. 1974). There are many examples of steroids that have no effect on fast feedback but produce inhibition some hours after administration (Jones and Tiptaft 1977). It is concluded, therefore, that the receptors for fast and delayed feedback must be different.

3 Fast Feedback in Man

Weitzman et al. (1971) demonstrated that cortisol was secreted episodically in man but they were unable to discover any constant relationship between plasma cortisol concentrations and the magnitude and timing of subsequent secretory episodes. From data obtained from infusion experiments with cortisol, Perlow et al. (1974) concluded that the attainment of plasma cortisol levels equal to those that occur spontaneously was unlikely to generate a negative feedback mechanism. Consequently there was considerable scepticism about the physiologic significance of feedback in man, and this was most strenuously expressed by James et al. (1978).

However, more recently, there have appeared several reports in the literature which amply confirm the existence of a rate-sensitive fast-feedback mechanism in man. Daly et al. (1979) found a negative correlation between plasma ACTH (determined by the cytochemical assay) and plasma cortisol, both at the trough and the peak of the circadian rhythm. This observation is compatible with, but does not prove, the existence of fast feedback. Subsequently, the same workers infused saline or cortisol at a steady rate of 3 mg $\cdot$ h^{-1} for 5 h. With cortisol, ACTH levels fell promptly and remained low for 2 h. Thereafter, the plasma ACTH rose to preinfusion levels. When the cortisol infusion rate was doubled to 6 mg $\cdot$ h^{-1} for a further 5 h there was a further dramatic drop in the ACTH levels which persisted for 3 h before returning to the preinfusion levels. On analysing the data it became apparent that the episodes of ACTH secretion were inhibited when the plasma cortisol levels were rising but there was a breakthrough when cortisol levels reached a plateau (Daly et al. 1979). These data support the possibility of a negative fast-feedback mechanism in man and suggest that the mechanism is sufficiently sensitive to be of physiologic significance. More recent data from the same laboratory is shown in Fig. 2. Here the episodic secretion of ACTH is abolished when the rate of change in plasma cortisol exceeds 5.1 nmol $\cdot$ 1^{-1} $\cdot$ min^{-1} (between 30–135 min) and 6.7 nmol $\cdot$ 1^{-1} $\cdot$ min^{-1} (between 345–405 min). It is also obvious from this figure that inhibition of ACTH secretion does not correlate with the absolute level of circulating cortisol nor with the total dose of cortisol that is administered. A further implication of these data is that the time domain for the emergence of delayed feedback is longer in man than that in the rat. In the former species it must be longer than 10 h but in the rat, the early phase of delayed feedback comes into operation at 60 min, or later. There is also some individual variation in the time of onset of fast feedback in man, there being a lag of 30 min before it occurred in the subject shown in Fig. 2, but there was only a delay of 15 min or less before it was observed in the published data on three normal individuals (Daly et al. 1979).

Recent work has demonstrated the existence of a negative fast-feedback (rate-sensitive) inhibition in patients with Addison's disease (Fehm et al. 1979) and in Cushing's disease (Carey 1980). Reader et al. (1980) investigated two patients with Cushing's disease (pituitary adenoma) and two patients with Cushing's syndrome (adrenal adenoma) and found that a rate of rise of 13.6–14.7 nmol $\cdot$ 1^{-1} $\cdot$ min^{-1} in plasma cortisol suppressed ACTH secretion in the patients with pituitary disease but a rise of only 1.3–1.6 nmol $\cdot$ 1^{-1} $\cdot$ min^{-1} was sufficient to depress ACTH to undetectable levels in the patients with adrenal tumors. They concluded that the fast feedback mechanism is more sensitive in patients with adrenal rather than pituitary adenomata.

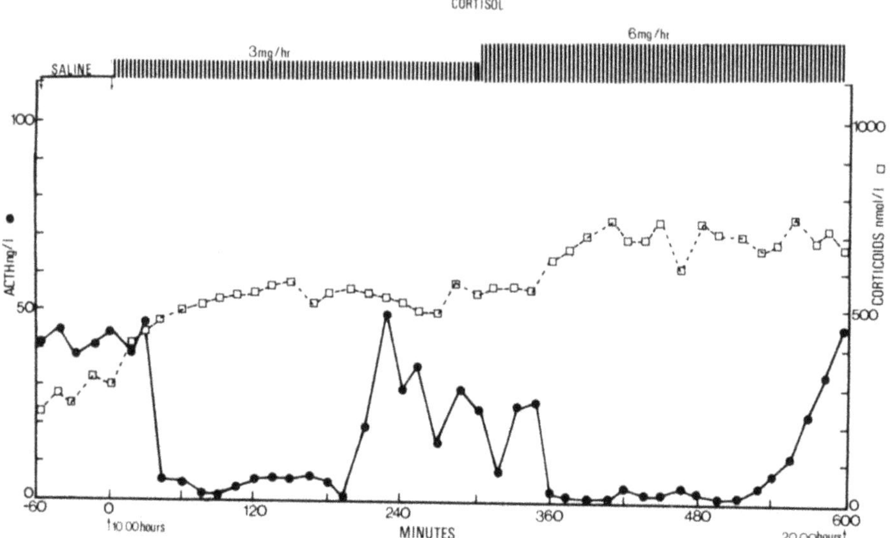

Fig. 2. Negative feedback inhibition of corticotrophin secretion by cortisol in the human. A normal human subject was infused with cortisol at 3 mg/h for 5 h, followed by 6 mg/h for a further 5 h. The cortisol infusion was preceded by a 1-h saline infusion, and plasma samples were taken at 15-min intervals and assayed for corticotrophin (by cytochemical bioassay) and corticosteroids (by fluorimetry).
Corticotrophin secretion fell abruptly when the corticosteroid level rose at an average rate of 5.1 (between 30 and 135 min) and 6.7 (between 345 and 405 min) nmols/liter per minute. Corticotrophin levels were below 5 ng/liter for the next 1.75 and 3 h respectively, after which secretion resumed during periods when corticosteroid levels remained steady. (Unpublished data, Reader, S.C.J., Robertson W.R., Alaghband-Zadeh, J., and Daly, J.R.)

Fehm (1977, 1979) found that in patients suffering from Addison's disease (adrenocortical hypofunction) the fast-feedback mechanism was demonstrably normal but in patients suffering from Cushing's disease there was a paradoxical rise in plasma ACTH concentration.

It would appear, therefore, that when the data produced from the same laboratory are compared (Daly et al. 1979; Reader et al. 1980) there is evidence for an abnormality of fast feedback in patients with Cushing's disease as compared with the response in normal individuals, i.e., the fast-feedback mechanism is blunted, in that something like twice the normal rate of rise in plasma cortisol is required to suppress ACTH secretion in Cushing's disease. Such an abnormality of the negative-feedback loop might be due to one of the basic mechanisms involved in the genesis of the disease. Fehm's data, with a paradoxical rise in plasma ACTH concentration, would be further evidence for an abnormality in the fast-feedback mechanism, albeit a more severe one where a normal negative response had been converted to a pathologic positive-feedback mechanism.

Whilst the information given above seems to show the existence of a fast-feedback mechanism in man the data are not directly comparable to those described in the rat.

In man the mechanism involved in the regulation of episodic (i.e., basal) secretion of ACTH was studied, whilst the work in the rat concerned the mechanism controlling the corticotrophic response to stress. There is no available evidence for the role of fast feedback in the control of the stress response in man. Nevertheless, it can be concluded that in man fast feedback is a physiologically relevant mechanism involved in the regulation of the basal secretion of ACTH although its role under stress conditions remains to be evaluated.

4 Delayed Feedback

It has become evident in recent years that much of the confusion in the literature concerning the effects of corticosteroids on the hypothalamo-pituitary-adrenocortical (HPA) axis arises from the failure to appreciate the importance of the dose and frequency of steroid administration, as well as the time at which the responsiveness of the system is tested.

It is clear that different effects are obtained depending upon whether the glucocorticoids are given as a single injection or part of a prolonged course.

4.1 Early Delayed Feedback

Many laboratories have investigated the effect of a wide variety of corticosteroids on the functional activity of the HPA axis 1–24 h following single or multiple injections of the steroids. On this basis some stresses were shown to be unaffected by corticosteroid treatment and were designated 'steroid-resistant' stresses whilst others were inhibited and were classified as 'steroid-sensitive' stresses (Dallman and Yates 1968). Dallman (1979) has recently reviewed such data from several laboratories and shown that in man, dog and rat there is good agreement across laboratories and between species as to stress classification (see Fig. 3). It appears, therefore, that some stresses (e.g., laparotomy plus gut traction or hemorrhage) are not affected by short-term treatment and truly activate steroid-resistant or high-threshold pathways.

Dallman (1979) has concluded, from the work in her own laboratory, and that of Sirett and Gibbs (1969), that if the treatment is repeated and prolonged to 16 or 28 h, with maximal doses of dexamethasone, these steroid resistant stresses become suppressed (Fig. 4). It is evident that repeated treatment accompanied by a prolonged time interval causes profound changes within the HPA axis. Indeed these effects are so different that it might be appropriate to think in terms of "early" and "late" delayed-feedback mechanisms. The recent findings by Herbert and co-authors (Roberts et al. 1979) provide evidence which throws some light on the possible biochemical basis for a difference between "early" and "late" delayed feedback. Herbert found that pretreatment with dexamethasone altered neither the concentration nor the rate of translation of mRNA for the ACTH precursor (pro-opiocortin) in mouse pituitary-derived At T-20 cells in culture until 6–12 h after treatment, the amount of message declining thereafter. It is therefore likely that the "steroid-resistant" stress-

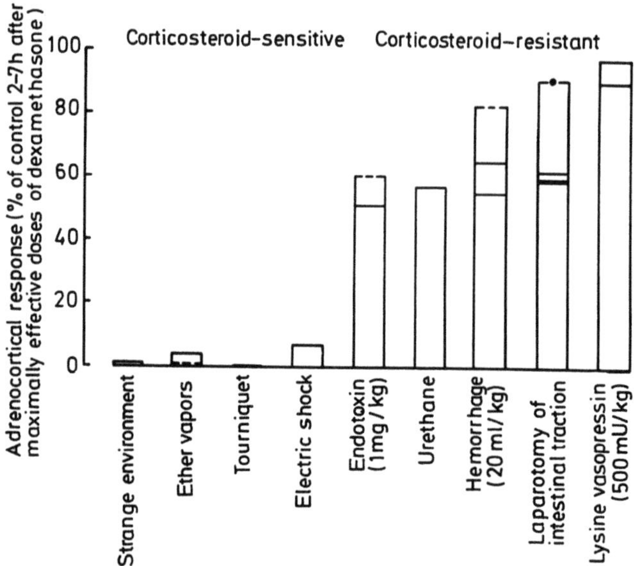

Fig. 3. Stimuli that normally activate the adrenocortical system are divided into those in which the adrenal response is blocked by pretreatment with a single large dose of corticosteroids (corticosteroid-sensitive) and those that still cause a response after pretreatment with a maximally effective dose of corticosteroids (corticosteroid-resistant). The figure is drawn from work from many laboratories and represents data obtained in man, dogs and rats. (Dallman 1979)

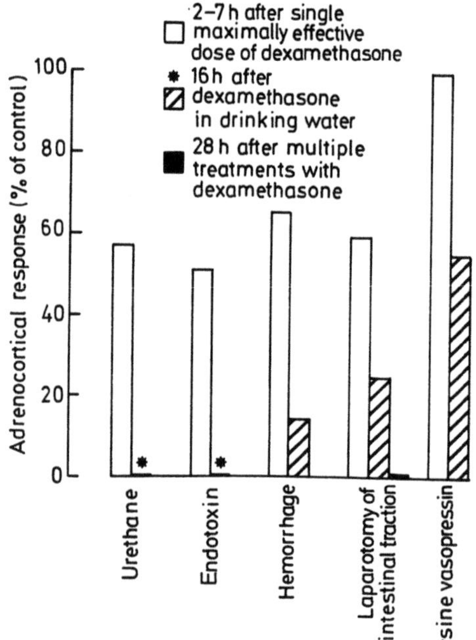

Fig. 4. The adrenal response to stimuli that after a single maximally effective dose of dexamethasone are progressively inhibited by prolonging the period of ingestion or injection of the steroid. The figure represents data redrawn from Sirett and Gibbs (1969, □ and ▨) and from Engeland et al. (1975, ▆). (Dallman 1979)

induced activation of the HPA axis becomes suppressible when the corticosteroids have prevented ACTH (and possibly CRF) synthesis. Doubtlessly stresses activate different neural pathways with varying degrees of steroid sensitivity but when the pituitary gland and hypothalamus fail to synthesize and release ACTH and CRF respectively the corticotrophic responses are suppressed

4.2 Late Delayed Feedback

This second phase of delayed feedback emerges when natural or synthetic glucocorticoids are administered for more than 12–24 h. The only stimulus to the axis which is unaffected is probably endotoxin (Stark et al. 1973/74), a stimulus which is capable of stimulating ACTH secretion in rats with isolated pituitaries (1974). Prolonged treatment with high doses of corticosteroids decreases the number of corticotrophic cells and the ACTH content of rat anterior pituitary glands (Siperstein and Miller 1970; Kraicer et al. 1973) and hyalin changes in ACTH-secreting cells in man (Halmi and Moriarty 1977). Stress in the rat causes an increase in the CRF activity of the median eminence and an increase in the ACTH content of the pituitary gland (Vernikos-Danellis 1963, 1965) and pretreatment with large doses of cortisol blocks both these effects. Similarly, the implantation of dexamethasone into the basal hypothalamus depresses, as tested 2 weeks later, CRF activity in the hypothalamus and the ACTH content in the pituitary gland (Chowers et al. 1967). Adrenal weight and adrenocortical responsiveness to ACTH are also decreased by 24 h (or later), after pretreatment with dexamethasone (Engeland et al. 1975; Dallman et al. 1978).

These data when viewed together show that chronic treatment with doses large enough to induce late delayed feedback abolishes activity of the axis in a hierarchical manner. The recovery of the axis appears to be in the same order with recovery of circulating ACTH levels to normal (or even supra normal) values before the return of adrenal responsiveness (Graber et al. 1965; Buckingham and Hodges 1976, 1977).

5 Sites of Action in Feedback Inhibition

5.1 Fast Feedback

5.1.1 Anterior Pituitary

Studies involving the use of dispersed pituitary cells in vitro have clearly shown the presence of an immediate feedback effect of corticosteroids (Sayers and Portanova 1974; Smelik 1977). Rats bearing lesions in the medial-basal hypothalamus (in which the release of endogeneous CRF is prevented) can also be used to demonstrate an immediate inhibitory effect of corticosteroids on the ACTH secretion induced by hypothalamic extracts or by CRF activity obtained by stimulating the rat hypothalamus in vitro with 5-hydroxytryptamine (Jones et al. 1979). When such rats are pretreated with corticosterone and the corticotrophic response to injection of hypothalamic extract is tested, it is found that the response is not inhibited at 5 min but is inhibited at 10, 20 and 30 min (Fig. 5). The measurement of plasma corticosterone in rats pretreated with corticosterone showed a peak concentration in the plasma at 10 min (Fig. 5). Thus inhibition at the pituitary level is seen when the concentration has reached a plateau and even when the plasma corticosterone is falling (i.e., 30 min) but not at the time when corticosterone is rapidly increasing (5 min). This suggests that fast feedback is not rate sensitive at this site, which is in contrast to observations on the corticotrophic response to stress in the normal animal where fast feedback shows rate sensitivity. It may therefore be concluded that the rate sensor is above the pituitary level.

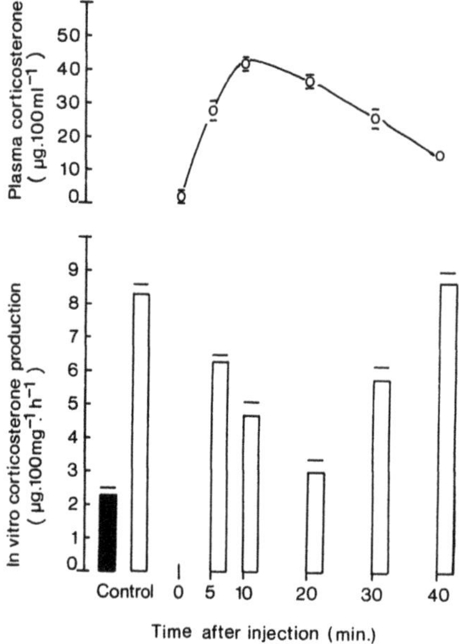

Fig. 5. The effect of injected corticosterone (100 µg/100 g s.c.) in basal hypothalamic lesioned rats on the corticotrophic response to hypothalamic extract (HE). The upper panel shows plasma corticosterone concentration and the lower panel the corticotrophic response as assessed by in vitro corticosterone production. HE was injected i.v. at 5–40 min after corticosterone administration. Control animals were injected with saline. O——O plasma corticosterone concentration; ■■ response to 0.5 ml saline (0.9% w/v) i.v.; ▭ response to 0.5 HE (1 hypothalamus/ml) i.v. (Mahmoud, Gillham and Jones, unpublished data)

The relevance of fast feedback at the anterior pituitary is doubtful, however, on other grounds. Thus there is a very poor correlation between the effect of steroids at the pituitary and their fast-feedback effects seen in vivo (Mahmoud et al., in preparation). Lastly, studies testing pituitary-adrenal responsiveness to a CRF preparation in vivo showed no effect of steroid fast-feedback effects on the response to CRF using a dose of hypothalamic extract that did not cause maximal pituitary activation (Leeman et al. 1962; Kaneko and Hiroshige 1978a).

5.1.2 Hypothalamus

Inhibition in vitro of neurotransmitter-induced CRF release by corticosterone and cortisol has been demonstrated using the rat hypothalamus (Jones et al. 1974; Jones and Hillhouse 1976) and hypothalamic synaptosomes (Edwardson and Bennett 1974). Results from in vivo studies support the conclusion that a rapid rate of rise in plasma corticosterone concentration inhibits CRF secretion when the steroid is given after application of a steroid-sensitive stress (Sato et al. 1975). Abe and Critchlow (1977) demonstrated fast feedback in the rat and found that the inhibition occurred when the hypothalamus was surgically isolated, thus showing that some feedback-mediating events must occur within the hypothalamo-pituitary complex.

5.1.3   Neural Pathways to the Hypothalamus

There is some evidence for an additional rate-sensitive component which is situated above the level of the hypothalamus. Thus Kaneko and Hiroshige (1978b) showed that there is normally an increase in hypothalamic CRF content following histamine stress which is prevented by prior treatment with corticosterone. However, after treatment with 6-hydroxydopamine this fast-feedback effect of corticosterone was lost, suggesting that the steroid required a noradrenergic neural input for its fast-feedback effect. It may be that there is more than one rate-sensitive system depending upon the pathways mediating the various stresses.

5.2   Early Delayed Feedback

5.2.1   Anterior Pituitary

A number of sites have been proposed for the delayed negative-feedback action of corticosteroids. These include the anterior pituitary gland, the hypothalamus, the amygdala and the hippocampus. Early observations tended to implicate the anterior pituitary gland. Thus Rose and Nelson (1956), by means of a micro-injection technique, were able to inhibit ACTH release by an infusion of cortisol of 20–37 μg/day into the pituitary fossa. Sayers and Portanova (1974) found that cortisol inhibited the CRF-induced release of ACTH from dispersed pituitary cells. ^{3}H-corticosterone is taken up in various sites in the rat brain, particularly the hippocampus, and dentate gyrus as well as the septum, and these sites are obvious candidates for feedback action. There is also uptake in the hypothalamus and the anterior pituitary but the amount, expressed on a weight basis, is much greater in the pituitary. As noted earlier, Vernikos-Danellis (1963, 1965) found that cortisol injections administered 4 h previously prevented the stress-induced increases in hypothalamic CRF as well as pituitary ACTH concentrations. These observations suggest hypothalamic as well as pituitary effects.

5.2.2   Hypothalamus

More direct evidence for early delayed feedback at the hypothalamus comes from experiments using the hypothalamus in vitro. Hypothalami were removed from rats pretreated 4 h previously with corticosteroids, and the ability of the tissue to produce CRF, in response to challenge with the excitatory neurotransmitters acetylcholine or 5-hydroxytryptamine was found to be reduced (Hillhouse and Jones 1976). More recently the early suppression of hypothalamic CRF activity by corticosteroids has been re-investigated in this laboratory. In these studies the determination of CRF activity was based on the ability of the releasing factor to stimulate ACTH secretion (determined cytochemically) by adenohypophysial segments in vitro. Figure 6 shows that dexamethasone (10^{-9} M) added to medium containing rat hypothalami pro-

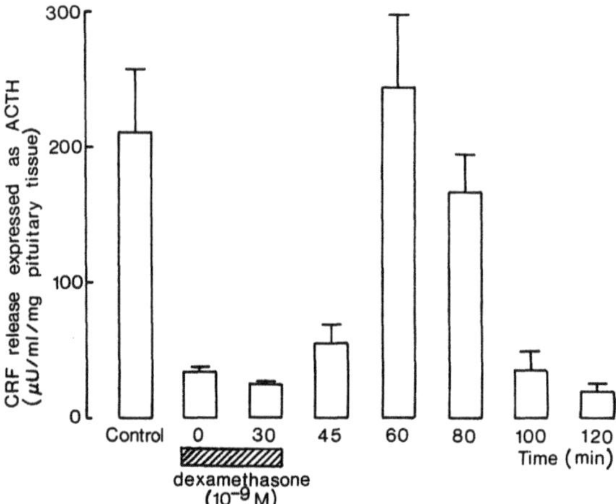

Fig. 6. Hypothalamic CRF release induced in vitro by acetylcholine (5 pg/ml) at various times after the addition and removal of dexamethasone from the incubation medium. Hypothalami were removed from male rats and incubated in an artificial medium resembling cerebrospinal fluid, for 30 min. Dexamethasone (10^{-9} M) was then added to the medium, and the hypothalami transferred, 30 min later, to medium alone for a further 90 min. The capacity of the tissue to release CRF was assessed at various times during the incubation. CRF activity was determined by measuring ACTH production (estimated cytochemically) induced by the releasing factor by adenohypophysial segments in vitro. Each value is the mean of five determinations ± SEM

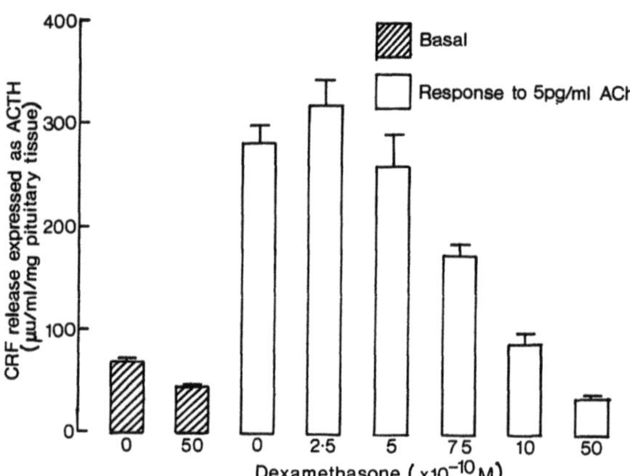

Fig. 7. Effect of dexamethasone on hypothalamic CRF activity induced in vitro by acetylcholine (ACh). Hypothalami were incubated with dexamethasone for 30 min and then in steroid free medium for a further 90 min, at which time their capacity to release CRF in response to the addition of ACh (5 pg/ml) was determined. CRF activity was measured as described in Fig. 6. Each value is the mean of five determinations ± SEM

duced an immediate inhibition of the acetylcholine-induced CRF release by the tissue, which lasted for 45 min. This was followed by a period between 60 and 80 min, when no inhibition was seen, before the onset of the second period of inhibition from 100 min onward. The extent of the second, delayed inhibition was shown to be dose related for concentrations of dexamethasone ranging from $5-50 \times 10^{10}$ M (Fig. 7). It has been shown previously that several steroids can antagonise the delayed effect of corticosterone on hypothalamic CRF activity in vitro (Jones and Hillhouse 1976). More detailed examination of the action of the antagonists revealed that prevention of the occurrence of the early delayed feedback can be achieved if the antagonist steroids are added up to 30 min after exposure of the tissue to the glucocorticoid (Fig. 8) (Beckford et al. 1981). It would appear from these data that there is a period after the addition of dexamethasone when the tissue may be 'rescued' from inhibition with the antagonist followed by a further period prior to the occurrence of delayed inhibition when rescue is not possible. The steroids that have these antagonistic properties are 11α-hydrocortisol, 17α- and 11α-hydroxyprogesterone and 11α, 17α-dihydroxy-progesterone. Whatever the nature of these irreversible events which make 'rescue' impossible, the final effect of the steroid is the inhibition of both the release and formation (presumably from a precursor) of CRF (Jones and Hillhouse 1977; Buckingham 1979).

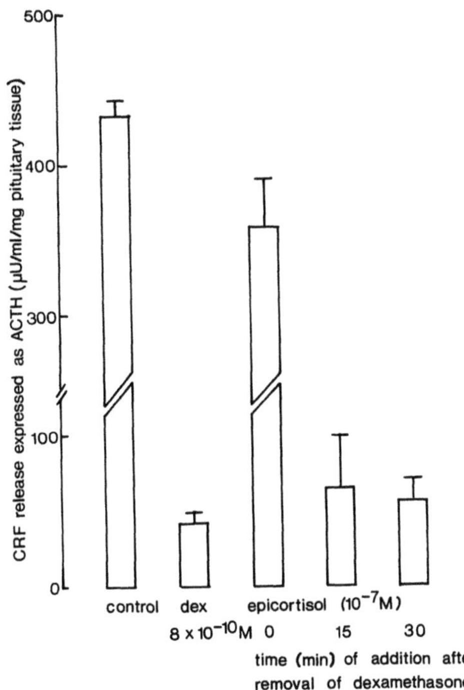

Fig. 8. Effect of epicortisol on the delayed inhibition induced in the hypothalamus in vitro by dexamethasone. Hypothalami were exposed to dexamethasone (8×10^{-10} M) for 30 min. The tissue was then transferred to medium containing epicortisol (10^{-7} M) (for 30 min) immediately after, 15 or 30 min after removal of dexamethasone. The capacity of hypothalami to release CRF in response to the addition of acetylcholine 5 pg/ml) was assessed 120 min from the time of addition of dexamethasone. CRF release was determined as described in Fig. 6. Each value is the mean of five determinations ± SEM

5.2.3 Neural Pathways to the Hypothalamus

There would appear to be little doubt that corticosteroids can act on neural pathways to the hypothalamus. Indeed, these may play a very important role in the modulation of the stress response induced via the early delayed feedback mechanism. This is indicated by the observations that whilst pretreatment of rats with cortisol or dexamethasone abolishes the normal increase in hypothalamic CRF activity after ether or sham adrenalectomy (Vernikos-Danellis 1965; Koch et al. 1975; Dallman and Yates 1968), even large doses of dexamethasone only partially reduce the increase in hypothalamic CRF content after laparotomy with intestinal traction (Takebe et al. 1971). These results are best understood if the former stresses act via steroid-sensitive pathways inhibiting input to the CRF cells whilst the latter stresses are mediated via steroid-resistant pathways. These observations do not exclude the possibility that the dexamethasone has an effect on cell bodies in the hypothalamus or on nerve terminals at this anatomical site, but it does promote the site in the neural hierarchy to somewhere above the CRF cells. The question which remains largely unanswered is the nature of the neurotransmitter pathway involved in this early delayed-feedback mechanisms. There is evidence in favor of two possible neurotransmitter mechanisms, gamma amino butyric acid (GABA) and serotonin (Acs and Stark 1978; Kaneko and Hiroshige 1978b).

5.3 Late Delayed Feedback

Whilst it may be true that neural pathways constitute an important site of early delayed feedback there appears little doubt that the important sites for late delayed feedback are exclusively the hypothalamus and the anterior pituitary. The principal evidence for this statement is that steroid-sensitive and steroid-insensitive stresses are inhibited following repeated treatment with large doses of glucocorticoids. In addition such treatment results in inhibition of CRF content and ACTH content (Dallman 1979).

These sites of late delayed feedback have also been attested to by the effect of chronic implantation of corticosteroids which have usually shown a greater potential for inhibiting ACTH secretion when they are placed in the hypothalamus rather than the pituitary (Smelik and Sawyer 1962; Chowers et al. 1963).

6 Mechanisms of Action

6.1 Fast Feedback

The immediacy of fast feedback is highly suggestive of an effect exerted at (or near to) the cell membrane. There is some circumstantial evidence to suggest such a membrane-mediated effect since the effect of glucocorticoids on CRF release is mimicked by manganese ions (which block Ca^{2+} flux) and antagonised by increasing the Ca^{2+} concentration in the incubating medium (Jones and Hillhouse 1976). The release of CRF has been shown to be a Ca^{2+}-dependent phenomenon (Jones and Hillhouse 1976;

Edwardson and Bennett 1974). Compatible with a membrane-mediated effect is the observation that fast feedback is associated with an inhibition of release but an increase in tissue content of CRF (Jones et al. 1976). Whilst the neurotransmitter-induced release of CRF (by either acetylcholine or serotonin) is inhibited by the addition of glucocorticoids, the release induced by prolonged depolarization by 48 mm K^+ or veratridine is unaffected (Jones and Hillhouse 1976; Smelik 1977). Both these stimuli would be expected to open wide the Ca^{2+} channels and this might be expected to swamp the putative inhibition of Ca^{2+} flux and exocytosis.

There are no reports in the literature on studies in which membrane binding of glucocorticoids (in the hypothalamus) has been investigated. There is clearly a need for such work to be done particularly since the structure-activity relationships of fast feedback have already been determined (Jones et al. 1977).

6.2 Early Delayed Feedback

Glucocorticoid binding occurs in several regions of the brain including the hypothalamus and also the anterior pituitary gland. In view of the existence of steroid-sensitive and steroid-insensitive neural pathways mediating various steroid-suppressible and steroid-insuppressible stresses, it is likely that the most important site of early delayed feedback is above the final common pathway of the CRF cells in the hypothalamus and the corticotrophic cells in the anterior pituitary gland. Since the nature of the pathways concerned have not been defined it is not possible to give an anatomic definition of the target sites. Some part of these pathways must inevitably terminate in the hypothalamus and it is conceivable that these endings themselves may be one of the target sites. There is certainly ample evidence for early delayed feedback at the hypothalamus and it is not surprising that this has been the subject of most of the studies. The hypothalamus may therefore have within its anatomic confines not only the CRF cells but also the associated axons terminating at the CRF neurons, all of which are possible targets of early delayed feedback.

Early delayed feedback is associated with inhibition of CRF release and formation (Jones and Hillhouse 1977; Buckingham 1979). The phenomenon has been observed to occur in vitro at 1 h or later after addition of the steroids. The question arises as to how much of this effect is genomic in nature. Early delayed feedback occurs within a time interval when it is possible for inhibition of mRNA synthesis to have occurred but it is questionable whether mRNA levels would have declined significantly by this time. A possible analogous phenomonen is what happens at the pituitary gland. Roberts et al. (1979) were able to infer the presence of specific receptors in the cytosol of cultured At T-20 cells. Addition of corticosteroids to the cells leads to the transfer of the receptor-steroid complex to the nucleus where the number of initiation sites for mRNA polymerase appears to be reduced. This results in a decline in the amount of mRNA for the ACTH precursor pro-opiocortin in the cells, and in pro-opiocortin itself. Although a similar model would be plausible for early delayed feedback at the hypothalamus, the detailed time courses of the events may be against this being the sole mechanism. As has been observed in vivo (Jones et al. 1974) and in vitro (Jones and Hillhouse 1976) delayed feedback can be observed 1–2 h after introduction of the steroid whereas in the work of Roberts et al. (1979) the first demonstrable fall in pro-opiocortin mRNA was not seen until after about 6 h, when the reduction was about 15% (pro-opiocortin had fallen less at that time). Thus although gene inactivation cannot be ruled out as mediating early delayed feedback it seems unlikely to explain the early inhibition. Indeed Roberts et al. (1979) found that inhibition of mRNA for pro-opiocortin was not maximal until 48 h after treatment. It would seem reasonable, therefore, to suggest the need for investigating the mechanism of early delayed feedback at the hypothalamus by looking at transcription and post-transcription — translation-related events.

The work of de Kloet et al. (1975) and McEwen (1977) has provided evidence for the specific high affinity binding of glucocorticoids in cytosolic preparations obtained from the rat hypothalamus, although this work tended to take attention away from the hypothalamus as a feedback site because the hippocampus was shown to have a higher concentration of specific, high affinity binding sites. The relevance of hippo-campal glucocorticoid binding to negative feedback regulation is most questionable, however, since these sites are normally saturated by resting levels of plasma cortico-sterone.

We have recently succeeded in confirming the existence of specific binding of ^{3}H-dexamethasone to cytosol binding sites, by adding the labeled steroid to the hypo-thalami in vitro followed by the preparation of a cytosolic fraction using two separate media, firstly a classical sucrose-phosphate medium and secondly a medium which we felt mimicked more closely the cationic composition of the intracellular medium. The composition of our intracellular-like medium is as follows: 80mM K_2HPO_4, 15mM NaH_2PO_4, 1mM $MgCl_2$.

The data obtained using the two media are shown in Fig. 9. It is evident that the number of binding sites is considerably greater using the intracellular-like medium but the K_d values are similar. The question arises as to what cation is mainly responsible for these effects. The inset illustrates the effect of changes in the K^+ concentration and shows that at levels of 50 mmol $\cdot$ 1^{-1} and higher the amount of bound steroid rises markedly (reaching a plateau at 90 mmol $\cdot$ 1^{-1}). It is interesting to note that based on calculations using the Nernst equation and using the published data on the resting membrane potential from hypothalamic slices (Mason 1980), the intraneuronal K^+ concentration for hypothalamic tissue would be 90mM. It is interesting that the cationic environment affects, in a similar fashion, the cytosol binding of labelled androgen (^{3}H-DHT) and this may have implications in other in vitro studies where experiments are carried out in what are cationically inappropriate environments.

It should be noted that when using the hypothalamus in vitro, the minimum effec-tive dose of dexamethasone which inhibits acetylcholine-induced release of CRF is very close to the K_d. This suggests that cytosolic binding may well be implicated in the events that inhibit CRF secretion via the early delayed feedback as tested 2 h after first exposure to the dexamethasone.

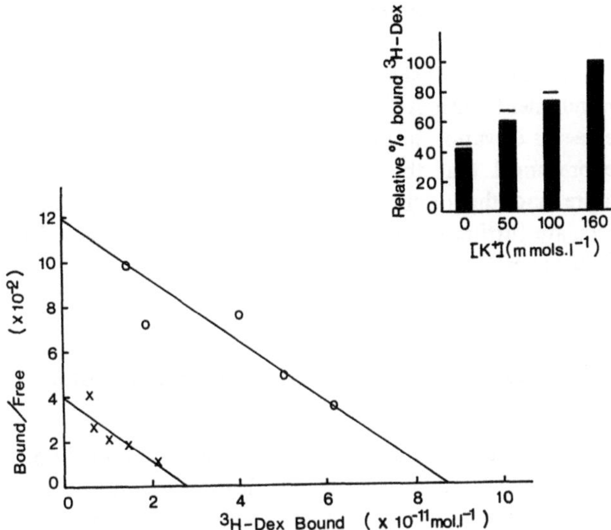

Fig. 9. Scatchard plots showing ^{3}H-dexamethasone binding sites extracted in sucrose-phosphate medium ($\times$) and a cationically intracellular-like medium (O). Hypothalami were removed from 7-day adrenalectomised male rats and incubated in vitro in a CSF-like medium with various concentrations of ^{3}H-dexamethasone (2×10^{-8} mol $\cdot$ 1^{-1} – 6.25×10^{-10} mol $\cdot$ 1^{-1}) in the presence and absence of excess cold dexamethasone for 2 h at 37°C. The tissue was then homogenised in either a sucrose-phosphate buffer containing $10mM$ Na Phosphate or a cationically intracellular-like medium. A cytosol fraction was prepared by centrifuging the homogenate at 105 000 g for 1 h at 4°C. Bound and free hormone were separated on columns of Sephadex LH-20 at 2°C. Radioactivity was measured by liquid scintillation spectrometry.
The inset shows the relative % bound ^{3}H-dexamethasone recovered after incubation of the hypothalami for 2 h with 10^{-8} mol $\cdot$ 1^{-1} ^{3}H-dexamethasone and homogenising the tissue in buffers containing increasing concentrations of K^+ ($0-160$ mmol $\cdot$ 1^{-1}). Columns represent mean $\pm$ SEM for six experiments.

6.3 Late Delayed Feedback

Chronic treatment with glucocorticoids reduces CRF and ACTH content in their respective tissues. From what has been stated earlier there appears little doubt that the effects of late delayed feedback are mediated by a reduction in mRNA for pro-opiocortin and possibly for CRF.

7 Feedback Regulation of Tissue CRF

Brodish (1977, 1979) has recently reviewed the work carried out (mainly in his own laboratory) on the regulaton of "tissue CRF", i.e., a humoral factor in peripheral blood that can activate the pituitary-adrenocortical axis. Tissue CRF differs in many ways from hypothalamic or median eminence CRF and its release is activated by extensive surgery, prolonged cold exposure and other stresses associated with tissue hypoxia. The origin of tissue CRF is unknown. However, there is clear evidence that it is not of hypothalamic origin. Nevertheless, its release appears to be inhibited by pretreatment with ACTH due to the elevation in plasma corticosterone levels. The degree of elevation in plasma corticosterone required is exceedingly high and it must be presumed that this negative-feedback mechanism has a high threshold. As the nature and the origin of this humoral factor is undetermined it is not possible to speculate further on the site or mechanism of action of the glucocorticoid inhibition of tissue CRF.

8   Feedback Effects of Various Glucocorticoids

It is often assumed that all glucocorticoids are equally effective at any of the various feedback sites. There is little justification for this assumption for it has been shown that the regional uptake and degree of nuclear uptake in different parts of the brain and anterior pituitary are not identical when a comparison is made between cortisol, corticosterone and the synthetic analogue dexamethasone (McEwen 1977). The differences in binding and efficacy at the anterior pituitary can be explained at least in part, by the presence of a transcortin-like binding protein which binds the natural occurring steroids but not dexamethasone which therefore has a readier access to the nuclei of the corticotroph cells (de Kloet et al. 1975; Koch et al. 1975, 1979). Vernikos-Danellis (1964) has inferred that hydrocortisone given 4 h previously inhibits stress-induced ACTH secretion even when the dose of the steroid is lower than that which is needed to block direct stimulation of the anterior pituitary gland by exogenous CRF. This suggests that the latter tissue is not the preferred target site for early delayed feedback.

There are some steroids which appear to inhibit at one site but not another. Thus, beclamethasone diproprionate has no discernable effect at the anterior pituitary but inhibits CRF release at the hypothalamus, whilst 17α-OH pregnenolone is active at the former but not the latter site (Jones et al. 1978).

It would appear, therefore, that there are significant differences in the affinity and efficacy of various steroids at the different feedback sites.

9   Integration of Fast and Delayed Feedback Mechanisms

An unanswered question concerns the relative roles and integration of the various feedback mechanisms in the control of ACTH secretion. Because of their long latency, neither early nor late delayed feedback mechanisms can regulate rapid responses, e.g., the stress-induced release of ACTH or its episodic secretion. It would appear at first sight that fast feedback alone could provide the necessary regulation of such rapid responses. Evidence for such a supposition comes from experiments conducted in rats, which were adrenalectomised and then immediately given a single injection of a small, physiologic dose of corticosterone. Circulating levels of ACTH were considerably elevated in adrenalectomised animals injected with saline but these levels were reduced to approximate toward those found in sham operated rats when there was replacement therapy with corticosterone (Dallman et al. 1972). However, this experiment was conducted under somewhat specialised circumstances where there would have been delayed-feedback signals provided by endogenous secretion during the hours preceding adrenalectomy. In other words, the experiments of Dallman et al. (1972) provide evidence for the efficacy of a fast-feedback mechanism when it operates against the background of adequate delayed-feedback signals.

Table 1. The interaction between fast- and delayed-feedback mechanism in the rat. All the animals used were adrenalectomized 3 days before the experiment, except for two groups that were sham operated at that time. Animals that received corticosterone before measurements of the basal or the stress-induced release of ACTH, were given acute steroid therapy (20 min before stress), or chronic steroid therapy (at 02.00 hours and 10.00 hours, last dose 2 h before stress), or a combination of both treatments.

Group	Treatment	Replacement category	Plasma ACTH (pg/ml)	
			Basal	Stress [a]
Sham operated	Saline	None	38.4 ± 3.8	247.5 ± 37.7
Adrenalectomized	Saline	None	277.4 ± 13.2	707.6 ± 99.0
Adrenalectomized or	Acute B (150 µg)	Acute	–	657.7 ± 56.2
	Acute B (550 µg)	Acute	–	524.5 ± 28.3
Adrenalectomized or	Chronic B (200 µg)	Chronic	75.8 ± 21.0	467.8 ± 95.8
	Chronic B (350 µg)	Chronic	–	476.3 ± 35.7
Adrenalectomized or	Acute B (150 µg)	Acute and	45.8 ± 9.6	273.5 ± 26.5
	Chronic B (200 µg)	chronic		

[a] Stress was caused by ether inhalation.

Further evidence for this interpretation comes from other experiments the results of which are depicted in Table 1. Rats were adrenalectomised and over the subsequent 3 days were given various forms of replacement therapy with corticosterone. Basal ACTH concentrations in the plasma as well as the corticotrophic response to ether stress were then compared to the values found in sham operated rats injected with saline. The purpose of the experiment was to determine what form of replacement therapy was necessary to normalize the ACTH levels to the values seen in the sham operated group but using a total dose of corticosterone which would be close to that secreted by the sham operated animals (approximately 500 µg). The effect of adrenalectomy is to cause an increase in basal and stress-induced release of radioimmunoassayable ACTH. Two delayed feedback signals of 200 µg (at 10.00 h and 20.00 h) over the 3 days followed by a fast-feedback signal 20 min before the stress (150 µg) resulted in normalization of both basal and stress-induced release of ACTH. When given alone, neither the fast-feedback signal (150 or 550 µg 20 min before stress) nor the chronic delayed-feedback signals (either 200 µg or 350 µg b.d.) were able to normalize the stress response. Thus both signals are needed for a normally responsive hypothalamo-pituitary axis. Of course, ACTH secretion can be normalized using higher doses of corticosterone (in the mg range) as delayed-feedback signals alone, but such doses are well in excess of the amounts of steroids the animals are able to secrete. This argues that the two feedback mechanisms constitute an integrated system that has evolved as an economical means of regulation in that it allows the stress response to occur and be modulated with amounts of corticosteroids that are insufficient to produce hyper-corticism. Following this line of argument, it is suggested that the delayed-feedback mechanism serves to reduce the responsiveness of the system to a level that can be modulated, on a moment to moment basis, by the fast-feedback mechanism. Examples of the operation of such moment to moment control are the episodic drive in the axis and the condition of acute stress.

The question which remains unresolved by these experiments is the relative importance of the early and late delayed-feedback mechanisms as a background enabling the fast feedback to operate effectively. Over a period of 3 days there would be sufficient time for the late delayed feedback to be in operation but the experiment was carried out 2 h after the last delayed-feedback signal, i.e., within the time domain of early phase. This problem can only be resolved by further experiments in which the effects of the two phases are examined separately. Since the mechanisms form an integrated regulatory system it would be of great interest to know the biochemical processes involved, and more particularly whether the degree of late delayed-feedback signal employed had influenced the level of mRNA for pro-opiocortin.

10 Species Variation

Although the occurrence of fast-feedback inhibition appears well documented in man and rats, it does not follow that such inhibition is found in all species. Thus, when Cowan and Windle (1978) gave two stepwise primed constant infusions of cortisol i.v. to adrenalectomized dogs and monitored the plasma ACTH concentration, they were unable to find evidence for a fast, rate-sensitive feedback inhibition of ACTH secretion. There are two methodological differences between the experiments and those performed in the rat, particularly in that there was no circulating cortisol in the dogs at the start of the experiments. However, in preliminary experiments Cowan has been unable to demonstrate fast feedback even when a prior delayed-feedback signal had been given (personal communication to MTJ).

It is of interest that in the dog the apparent absence of a fast-feedback inhibitory system appears to be accompanied by an early delayed-feedback system that can be activated as quickly as 20 min after steroid infusion (Cowan and Windle 1978). This may indicate that the time domain for early delayed feedback may vary from species to species.

11 Conclusion

To date, most experimental investigations of glucocorticoid-mediated inhibition of adrenocortical activity have been directed toward the anterior pituitary gland and the hypothalamus simply because it is relatively easy to manipulate the concentration of corticosterone (and/or cortisol) and to measure an end point, viz ACTH secretion. Of course, when these studies are at the hypothalamic level the question of the nature of CRF must cloud any interpretation of results. It is for this reason that data obtained in vitro must always be checked by experiment in vivo.

Now that techniques are becoming more refined, it is appropriate to commence a fuller investigation of possible feedback events arising in the brain in extrahypothalamic regions, when results might be expected to complement those already obtained from studies at the anterior pituitary and hypothalamus.

References

Abe K, Critchlow V (1977) Effects of corticosterone, dexamethasone and surgical isolation of the medial basal hypothalamus on rapid-feedback control of stress induced corticotrophin-secretion in female rats. Endocrinology 101:498–505

Acs Z, Stark E (1978) Possible role of gamma-amino-butyric acid synthesis in the mechanism of dexamethasone feedback action. J Endocrinol 77:137–141

Beckford U, Gillham B, Greenstein BD, Holmes MC, Jones MT (1981) J Physiology 316:39–40

Bethune JE, Nelson DH, Thorn GW (1957) Plasma adrenocorticotrophic hormone in Addison's disease and its modification by the administration of adrenal steroid. J Clin Invest 36:1701–1707

Brodish A (1977) Extra-CNS Corticotrophin-releasing factors. Annals NY Acad Sci 297:420–432

Brodish A (1979) Control of ACTH secretion by corticotropin releasing factor(s). Vitam Horm 37:111–152

Buckingham JC (1979) The influence of corticosteroids on the secretion of corticotrophin and its hypothalamic releasing hormone. J Physiol 286:331–342

Buckingham JC, Hodges JR (1976) Hypothalamo-pituitary adrenocortical function in the rat after treatment with betamethasone. Br J Pharmacol 56:235–239

Buckingham JC, Hodges JR (1977) Functional activity of the hypothalamo-pituitary complex in the rat after treatment with betamethasone. J Endocrinol 74:297–302

Carey RM (1980) Suppression of ACTH by cortisol in dexamethasone-non suppressible Cushing's disease. N Engl J Med 302:275–279

Chowers I, Feldman S, Davidson JM (1963) Effects of intrahypothalamic crystalline steroids on acute ACTH secretion. Am J Physiol 205:671–673

Chowers I, Conforti N, Feldman S (1967) Effects of corticosteroids on hypothalamic corticotropin releasing factor and pituitary ACTH content. Neuroendocrinology 2:193–199

Cowan JS, Windle JS (1978) Progressive suppression of adrenocorticotropin secretion in resting adrenalectomised dogs by low stepwise infusions of cortisol. Endorinology 103:1173–1182

Dallman MF (1979) Adrenal feedback on stress-induced corticoliberin (CRF) and corticotropin (ACTH) secretion. In: Jones MT, Gillham B, Dallman MF, Chattopadhyay S (eds) Interaction within the brain-pituitary-adrenocortical system. Academic Press, London, pp 149–162

Dallman MF, Jones MT (1973) Corticosteroid feedback control of ACTH secretion: effects of stress-induced corticosterone secretion on subsequent stress response in the rat. Endocrinology 92:1367–1375

Dallman MF, Yates FE (1968) Anatomical and functional mapping of central neural input and feedback pathways of the adrenocortical system. Mem Soc Endocrinol 17:39–72

Dallman MF, Yates FE (1969) Dynamic asymmetries in the corticosteroid feedback path and distribution-metabolism-binding-elements of the adrenocortical system. Ann NY Acad Sci 156(2):696–721

Dallman MF, Jones MT, Vernikos-Danellis J, Ganong WF (1972) Corticosteroid feedback control of ACTH secretion: Rapid effects of bilateral adrenalectomy on plasma ACTH in the rat. Endocrinology 91:961–968

Dallman MF, Engeland WC, Rose JC, Wilkinson CW, Shinsako J, Siedenburg F (1978) Nycthemeral rhythm in adrenal responsiveness to ACTH. Am J Physiol 235: R210–218

Daly JR, Fleishner MR, Chambers DJ, Bitensky L, Chayen J (1974) Application of the cytochemical bioassay for corticotrophin to clinical and physiological studies in man. Clin Endocrinol 3:335–345

Daly JR, Reader SCJ, Alaghband-Zadeh J, Halsmann P (1979) Observations on feedback regulation of corticotropin (ACTH) secretion in man. In: Jones MT, Gillham B, Dallman MF, Chattopadhyay S (eds) Interaction within the brain-pituitary-adrenocortical system. Academic Press, London, pp 181–188

Deane HW, Greep RO (1946) A morphological and histochemical study of the rat's adrenal cortex after hypophysectomy, with comments on the liver. Am J Anat 79: 117–146

Edwardson JA, Bennett GW (1974) Modulation of CRF release from hypothalamic synaptosomes. Nature (Lond) 251:425–427

Engeland WC, Shinsako J, Dallman MF (1975) Corticosteroids and ACTH are not required for compensatory adrenal growth. Am J Physiol 229:1461–1464

Fehm HL, Voigt KH, Lang RE, Beinert KE, Kummer GW, Pfeiffer EF (1977) Paradoxical ACTH response to glucocorticoids in Cushing's disease. N Engl J Med 297:904–907

Fehm HL, Voigt KH, Kummer G, Lang R, Pfeiffer EF (1979) Differential and integral corticosteroid feedback effects on ACTH secretion in hypoadrenocorticism. J Clin Invest 63:247–253

Graber AL, Ney RL, Nicholson WE, Island DP, and Liddle GW (1965) Natural history of pituitary-adrenal recovery following long-term suppression with corticosteroids. J Clin Endocrinol Metab 25:11–16

Halmi NS, Moriarty GC (1977) The cells of origin of ACTH in man. Ann NY Acad Sci 297:167–181

Hillhouse EW, Jones MT (1976) Effect of bilateral adrenalectomy and corticosteroid therapy on the secretion of corticotrophin-releasing factor activity from the hypothalamus of the rat in vitro. J Endocrinol 71:21–30

Hodges JR, Jones MT (1963) The effect of injected corticosterone on the release of ACTH in rats exposed to acute stress. J Physiol (Lond) 167:30–37

Ingle DJ, Kendall EC (1937) Atrophy of the adrenal cortex in the rat produced by the administration of large amounts of cortin. Science NY 86:245–252

Ingle DJ, Higgins GM, Kendall EC (1938) Atrophy of the adrenal cortex of the rat produced by administration of large amounts of cortin. Anat Rec 71:363–372

James VHT, Landon J, Fraser R (1968) Some observations on the control of corticosteroid secretion in man. Mem Soc Endocrinol 17:141–158

James VHT, Tunbridge RDG, Wilson GA, Hutton J, Jacobs HS, Rippon AE (1978) Steroid profiling: A technique for exploring adrenocortical physiology. In: James VHT, Serio M, Giusti G, Martini L (eds) The endocrine function of the human adrenal cortex. Academic Press, London, p 179

Jones MT, Hillhouse EW (1976) Structure-activity relationship and the mode of action of corticosteroid feedback on the secretion of corticotrophin-releasing factor (corticoliberin). J Steroid Biochem 7:1189–1202

Jones MT, Hillhouse EW (1977) Neurotransmitter regulation of corticotrophin releasing factor in vitro. Ann NY Acad Sci 297:536–560

Jones MT, Tiptaft EM (1977) Structure activity relationship of various corticosteroids on the feedback control of corticotrophin secretion. Br J Pharmacol 59:35–41

Jones MT, Brush FR, Neame RLB (1972) Characterisitcs of fast feedback control of corticotrophin release by corticosteroids. J Endocrinol 55:489–497

Jones MT, Tiptaft EM, Brush FR, Fergusson DAN, Neame RLB (1974) Evidence for dual corticosteroid receptor mechanisms in the feedback control of ACTH secretion. J Endocrinol 60:223–233

Jones MT, Hillhouse EW, Burden JL (1976) Secretion of corticotropin releasing hormone in vitro. Front Neuroendocrinol 4:195–226

Jones MT, Hillhouse EW, Burden JL (1977) Structure activity relationships of corticosteroid feedback at the hypothalamic level. J Endocrinol 74:415–424

Jones MT, Gillham B, Mahmoud S (1978) Hypothalamus and ACTH secretion. In: James VHT, Serio M, Giusti G, Martini L (eds) The endocrine function of the human adrenal cortex. Academic Press, London, pp 55–85

Jones MT, Gillham B, Mahmoud S, Holmes MC (1979) The characteristics and mechanism of action of corticosteroid negative feedback at the hypothalamus and anterior pituitary. In: Jones MT, Gillham B, Dallman MF, Chattopadhyay S (eds) Interaction within the brain-pituitary-adrenocortical system. Academic Press, London, pp 163–180

Kaneko M, Hiroshige T (1976) Quantitative analysis of rate sensitive negative feedback regulation of ACTH secretion under stress in the rat. (Abstract). V International Congress Endocrinology, Hamburg, pp 132

Kaneko M, Hiroshige T (1978a) Fast, rate-sensitive corticosteroid negative feedback during stress. Am J Physiol 234:R39–R45

Kaneko M, Hiroshige T (1978b) Site of fast, rate-sensitive feedback inhibition of adrenocorticotrophin secretion during stress. Am J Physiol 234:R46–R51

de Kloet R, Wallach G, McEwen BS (1975) Differences in corticosterone and dexamethasone binding to rat brain and pituitary. Endocrinology 96:598–609

Koch B, Lutz B, Briand B, Mialhe C (1975) Glucocorticoid binding to adenohypophysis receptors and its physiological role. Neuroendocrinology 18:299–310

Koch B, Lutz-Bucher B, Briand B, Mialhe C (1979) Relationship between ACTH secretion and corticoid binding to specific receptors in perifused adenohypophyses. Neuroendocrinology 28:169–177

Kraicer J, Gosbee JL, Bencosme SA (1973) Pars intermedia and pars distalis: two sites of ACTH production in the rat hypophysis. Neuroendocrinology 11:156–176

Leeman SE, Glenister DW, Yates FE (1962) Characterization of a calf hypothalamic extract with adrenocorticotropin-releasing properties: evidence for a central nervous system site for corticosteroid inhibition of adrenocorticotropin release. Endocrinology 70:249–262

McEwen BS (1977) Adrenal steroid feedback on neuroendocrine tissues. Ann NY Acad Sci 297:568–579

Mason WT (1980) Supraoptic neurones of rat hypothalamus are osmosensitive. Nature (Lond) 287:154–157

Perlow M, Weitzman ED, Hellman L (1974) Effect of cortisol infusions on endogenous cortisol secretion in man. Endocrinology 95:790–795

Reader SCJ, Robertson WR, Alaghband-Zadeh J, Daly JR (1980) Negative feedback effects on adrenocorticotrophin secretion by cortisol in Cushing's syndrome. J Endocr 87:60–61P

Roberts JL, Budarf ML, Baxter JD, Herbert E (1979) Selective reduction of proadrenocorticotrophin/endorphin proteins and messenger ribonucleic acid activity in mouse pituitary tumor cells by glucocorticoids. Biochemistry 18:4907–4915

Rose S, Nelson J (1956) Hydrocortisone and ACTH release. Aus J Exp Biol Med Sci 34:77–82

Sato T, Sato M, Shinsako J, Dallman MF (1975) Corticosterone-induced changes in hypothalamic corticotropin-releasing factor (CRF) content after stress. Endocrinology 97:265–274

Sayers G, Portanova R (1974) Secretion of ACTH by isolated anterior pituitary cells: kinetics of stimulation by CRF and inhibition by corticosterone. Fed Proc 32:295

Sayers G, Sayers MA (1948) Regulation of pituitary adrenocorticotrophic activity during the response of the rat to acute stress. Endocrinology 40:265–273

Siperstein ER, Miller KJS (1970) Further cytophysiologic evidence for the cells that produce adrenocorticotrophic hormone. Endocrinology 86:451–486

Sirett NE, Gibbs FP (1969) Dexamethasone suppression of ACTH release: effect of the interval between steroid administration and the application of stimuli known to release ACTH. Endocrinology 85:355–359

Smelik PG (1963) Failure to inhibit corticotrophin secretion by experimentally induced increases in corticoid levels. Acta Endocrinol (Kbh) 44:36–46

Smelik PB (1977) Some aspects of corticosteroid feedback actions. Ann NY Acad Sci 297:580–588

Smelik PG, Sawyer CH (1962) Effects of implantation of cortisol into the brainstem or pituitary gland on the adrenal response to stress in the rabbit. Acta Endocrinol (Kbh) 41:561–570

Stark E, Makara GB, Marron J, Palkovits M (1973/74) ACTH release in rats after removal of the medial hypothalamus. Neuroendocrinology 13:224–233

Takebe K, Kunita H, Sakakura M, Horiuchi Y, Mashimo K (1971) Suppressive effect of dexamethasone on the rise of CRF activity in the median eminence induced by stress. Endocrinology 89:1014–1019

Vernikos-Danellis J (1963) Effect of acute stress on the pituitary gland: Changes in blood and pituitary ACTH concentrations. Endocrinology 72:574–581

Vernikos-Danellis J (1964) Estimation of corticotropin-releasing activity of rat hypothalamus and neurohypophysis before and after stress. Endocrinology 75:514–520

Vernikos-Danellis J (1965) Effect of stress, adrenalectomy, hypophysectomy and hydrocortisone on the CRF activity of the rat median eminence. Endocrinology 76:122–126

Weitzman ED, Fukushima D, Nogeire C, Roffwarg H, Gallagher TF, Hellman L (1971) Twenty-four hour pattern of the episodic secretion of cortisol in normal human subjects. J Clin Endocrinol Metab 33:14–22

Winter CA, Silver RH, Stoerk HC (1950) Production of reversible hyperadrenocortinism in rats by prolonged administration of cortisone. Endocrinology 47:60–72

Yates FE, Leeman SE, Glenister DW, Dallman MF (1961) Interaction between plasma corticosterone concentration and ACTH-releasing stimuli in the rat: Evidence for the reset of an endocrine feedback control. Endocrinology 69:67–80

Electrophysiologic and Clinical Aspects of Glucocorticoids on Certain Neural Systems

WALTER F. RIKER, Jr.[1], THOMAS BAKER[2], and ANTONIO SASTRE[3]

Contents

1 Purpose

This review considers the premise that certain neural effects of exogenously administered glucocorticoids arise from direct, non-receptor-mediated actions. It is emphasized that these effects are not regarded as alternatives to the well-documented actions mediated via cytoplasmic receptors and nuclear actions (see chapters by McEwen and

---

1 Revlon Professor and Chairman, Department of Pharmacology, Cornell University Medical College, 1300 York Avenue, New York, NY 10021/USA
2 Associate Research Professor, Department of Pharmacology, Cornell University Medical College, 1300 York Avenue, New York, NY 10021/USA
3 Assistant Professor, Department of Physiology and Neuroscience, Johns Hopkins School of Medicine, 725 North Wolfe Street, Baltimore, MD 21205/USA

Bohus); rather, these effects are considered in addition to classic mechanisms. In a concluding section, data are presented to support the concept that these glucocorticoid effects result from a direct action on susceptible membranes. The evidence discussed suggests that particular actions of exogenously administered glucocorticoids, especially with repeated high dosage, cause changes in the electrical excitability of certain susceptible neuronal structures and systems. These actions undoubtedly occur pari passu with the well-known cytoplasmic and nuclear events of these steroid hormones.

The skeletal neuromuscular junction is a focal point for this review, chiefly because it is an accessible model synapse in which these presumably direct glucocorticoid effects have been manifested and described. Comparable sites of glucocorticoid action in the spinal cord and brain are also considered. Particular attention is given to experiments in which the time course of the response (in the order of microseconds to seconds) or the experimental circumstances make it unlikely that the glucocorticoid-induced effect is mediated via a cytoplasmic receptor mechanism. Most of the experiments reviewed here are pharmacologic in their nature and implications.

2 Background

Among the steroid hormones, the effects of glucocorticoids on nervous system excitability are exceptional. This was revealed initially by Woodbury and his co-workers (Woodbury 1952; cf. Woodbury 1958; Timiras et al. 1956; Woodbury and Vernadakis 1966; Withrow and Woodbury 1972). They tested nervous system excitability in rats by means of electroshock seizure threshold (EST), a gross but reliable assay. They found that chronic intensive dosing with cortisol and cortisone (2 mg/kg i.m. daily for 4–28 days) greatly enhances brain excitability. Holtkamp et al. (1952), using cortisone, described the dose dependency of this effect. It was soon learned that the same increase in brain excitability in rats is achievable with the synthetic glucocorticoid, prednisone (Mansor et al. 1956).

The mineralocorticoids, desoxycorticosterone and 11-deoxycortisol, when used at similar doses, lessened brain excitability in rats. Withrow and Woodbury (1964, 1972) later attribtued excitability changes following mineralocorticoid excess or deficiency to severe intra- and extracellular electrolyte alterations. The sex hormones, testosterone and estradiol, alter neuronal excitability in a biphasic fashion in discrete brain areas (e.g., hippocampus and preoptic nuclei). The effect depends on the sex of the animal and the stage of the estrous cycle (Kelly et al. 1977; Teyler et al. 1980). A direct depressant effect of progesterone was found in man and led, via structure-activity studies,

---

The following abbreviations appear in the text:

ACh: acetylcholine; ACTH: adrenocorticotropic hormone; CNS: central nervous system; dTC: d-tubocurarine; EEG: electroencephalograph(ic); EST: electroshock seizure threshold; HACA: high-affinity choline accumulation; HC-3: hemicholinium-3; mepp: miniature endplate potential; MSR: monosynaptic reflex pathway; PTP: posttetanic potentiation; SBR: stimulus-bound repetition; TTX: tetrodotoxin

to the development of general anesthetic drugs, e.g., hydroxydione, 5β-pregnane-3,20-dione (P'an and Laubach 1964). These steroids lack hormonal activity. Although these drugs have not gained clinical acceptance as anesthetics, their neurodepressant activity shows that some steroids can affect neural excitability and that this direct neural action is separable from hormonal action.

In early work, spontaneous seizures were seen to occur in animals and patients treated with high doses of adrenocorticotropic hormone (ACTH) or cortisone (Astwood et al. 1950a, b; Dameshek et al. 1950; Dorfman et al. 1951; Pincus et al. 1951; Stephen and Noad 1951; Hicks 1953). Glaser et al. (1955) correlated the laboratory and clinical experiences by showing that intravenous cortisol in humans increases the frequency of the electroencephalographic (EEG) alpha rhythm. He had previously reported similar EEG changes and seizure episodes in patients with Cushing's disease (Glaser 1953).

Early research reports concurred with these observations. Cortisone and ACTH decreased seizure threshold in dogs (Pasolini 1952). Cortisone promoted the cortical irradiation of evoked responses in rabbits (Misrahy and Toman 1953). Heightened brain excitability in rats given high doses of cortisol was further evidenced by a prolongation of the tonic component of the seizures and a shortened recovery time, i.e., a reduced postictal depression (Timiras et al. 1956).

Shortly after these early disclosures of a gross excitatory effect of glucocorticoids on brain, a few investigations revealed that glucocorticoids increase excitability in particular CNS pathways. Feldman et al. (1961) and Endroczi and Koranyi (1968) demonstrated that cortisol, given intravenously to cats, increases the amplitudes of sensory evoked potentials in the brain stem reticular formation, in certain hypothalamic areas, and in intralaminar thalamic nuclei. Similarly, Koranyi and Endroczi (1970) showed that cortisol augments the diffusely projected thalamic recruiting response. Recordings in the lemniscal path showed no changes and these authors (Feldman et al. 1961; Endroczi and Koranyi 1968) concluded that multisynaptic pathways are especially susceptible. However, cortisol has also been shown to amplify evoked potentials in oligosynaptic sensory pathways (Hoagland et al. 1953; Covian et al. 1963; Koranyi and Endroczi 1970).

Recent investigations of glucocorticoid effect on motor nerve excitabilities (see Sect. 3.1) and single-unit activity in brain show that polysynaptic connectivity is not a requirement for this excitability enhancement. Importantly, however, these more recent findings also reveal that glucocorticoids may cause opposite excitability changes in otherwise closely related neuron types, as well as biphasic effects in the same neuron (vide infra).

The results of Hoagland et al. (1953), Feldman et al. (1961), Covian et al. (1963), Endroczi and Koranyi (1968), and Koranyi and Endroczi (1970) showed that certain diencephalic and hypothalamic neurons respond biphasically to glucocorticoids. For example, studies designed to detect diencephalic loci of hormonal feedback (Davidson and Feldman 1963; Mangili et al. 1966; Slusher et al. 1966; Feldman and Dafny 1970a, b; Steiner 1970; Pfaff et al. 1971) report that spontaneous activity in sampled neurons in various brain regions may be either accelerated or inhibited by both cortisol and dexamethasone. Sequential inhibition and facilitation of spontaneously discharging snail neurons also occurred after iontophoretic application of dexamethasone (Schade and van Wilgenburg 1973). These biphasic effects resemble those described later on for motor nerves.

More recently, a single-unit study by Barak et al. (1977) revealed a degree of neuronal selectivity for the direct effects of glucocorticoids on excitability. Using pressure-injection and iontophoretic techniques, these authors observed a reduction in spontaneous firing of neurons in the mediobasal hypothalamus of rats and cats on application of corticosterone and cortisol. These effects took place within 1–2 s, suggesting a direct membrane effect. In contrast, the same iontophoretic pipets failed to affect the firing of over 400 hippocampal cells tested. These experiments, together with others to be described, establish that glucocorticoids can selectively affect different neuronal subtypes.

In search of mechanisms to explain steroid effects on whole brain excitability, Woodbury (1954, 1972), Timiras et al. (1954) and Withrow and Woodbury (1972) measured brain electrolytes. Although the effects of desoxycorticosterone on excitability could be related to increases in brain intracellular Na^+ concentrations, the glucocorticoid effect was not so clearly correlated. Woodbury and Vernadakis (1966) and Withrow and Woodbury (1972) suggested that cortisol augments brain excitability by increasing brain cell permeability to Na^+. No characteristic metabolic effect of glucocorticoids could be correlated with this excitability enhancement (Woodbury and Vernadakis 1966). There has been little work since these early efforts to explore other possible mechanisms by which glucocorticoids might increase excitability in neural systems. Recent studies of glucocorticoid effects on motor nerve and neuromuscular transmission, in this and other laboratories, have led to the suggestion that direct effects on certain excitable membranes may account for the observed excitability changes (Riker et al. 1975; Baker et al. 1977; Hall 1980a).

This background statement can be closed appropriately with a quote from the chapter on adrenocortical steroids in the sixth edition (1980) of Goodman and Gilman's classic text: "The corticosteroids affect the central nervous system in a number of indirect ways . . . The steroids may also have direct effects, but these are as yet poorly defined."

3 Glucocorticoid Effects on Cat Motor Nerve Terminals In Vivo

3.1 Introduction

Laboratory studies of glucocorticoid effects on transmitter function have been pursued in recent years mostly with neuromuscular preparations. This direction was largely motivated by the effectiveness of glucocorticoids in the treatment of myasthenia gravis, as discussed further on in this chapter, and by the question of whether this steroid effect might involve direct actions on neuromuscular structures.

These studies disclosed that glucocorticoids exert significant actions on motor nerve terminals (Patten et al. 1974; Wilson et al. 1974; Riker et al. 1975). Other endeavors have since confirmed this (see Sect. 7). Therefore, this review focuses on glucocorticoid effects on neuromuscular function with particular reference to motor nerve terminals.

Because extensive studies of motor nerve terminal pharmacology have been conducted with cat neuromuscular preparations (Standaert and Riker 1967; Riker and Okamoto 1969; Riker 1975; Miyamoto 1977; Bowman and Rand 1980), much of the work with glucocorticoids has involved cat in vivo preparations. This may prove clinically relevant in that neuromuscular function and pharmacology in the cat are remarkably like those entities in humans.

3.2 Methodology

Glucocorticoid effects on motor nerve function in vivo have been identified in one way by means of a cat soleus nerve-muscle preparation (Riker et al. 1975; Baker et al. 1977; Hall et al. 1977a). The principles of this method are briefly described here because the technique is not widely used and an appreciation of the mechanisms represented aids in the understanding of the findings. The experimental method is sketched in Fig. 1.

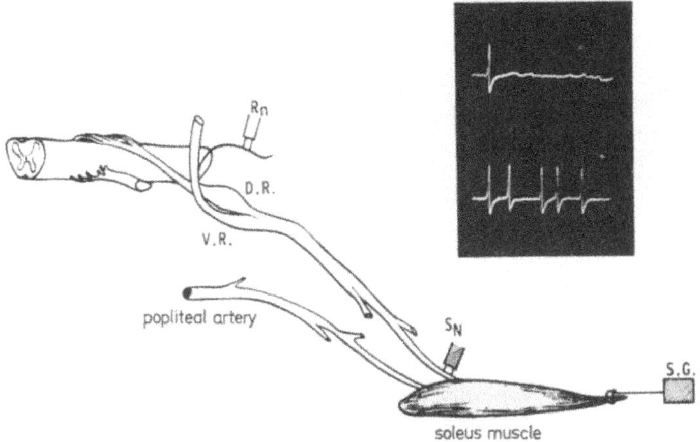

Fig. 1. Scheme of the method for recording antidromic neural activity. Stimuli are applied to the peripheral nerve with electrode S_N. Electrode R_n records from a ventral root filament containing a single active motor axon. *D.R.*, dorsal root; *V.R.*, ventral root; *S.G.*, strain gauge. *Inset:* The *upper tracing* is a control; the *lower* shows the neural repetitive response to a single stimulus following preconditioning of the nerve terminals. (Riker WF, Standaert FG 1966)

Brief high-frequency impulse trains traversing motor nerve terminals facilitate their function; transmitter mobilization is increased and transmitter release facilitated (Liley and North 1953; Hubbard and Schmidt 1963; Volle 1966; Maeno 1969; Maeno and Edwards 1969; Zucker 1973). Cat soleus motor nerve terminals manifest a frequency-induced facilitation in an unusual, but remarkably useful, way (Standaert 1963, 1964). Following an adequate high-frequency impulse barrage (200—400 Hz for 10 s), these terminals apparently hyperpolarize. Like hyperpolarization in other small-diameter unmyelinated nerves, each subsequent single excitation can be expected to evoke a longer-lasting action potential than in normal terminals (Werner 1960; Standaert 1963; Hubbard and Schmidt 1963; Katz and Miledi 1965; Hubbard et al. 1965). This prolonged activity in the terminals serves to reexcite repetitively the contiguous (and much more quickly recovered) parent axon (Fig. 1, inset). Because this repetitive excitation of motor axons occurs after only a single excitation of the facilitated terminals, it is called a stimulus-bound repetition (SBR). A scheme of this mechanism is illustrated in Fig. 2. Because SBR is propagated both ortho- and antidromically, it is readily monitored in both nerve and muscle. Monitoring the antidromic SBR from ventral root portions of motor axons enables a direct identification of motor nerve terminal activity because transmission is not involved (see Fig. 1).

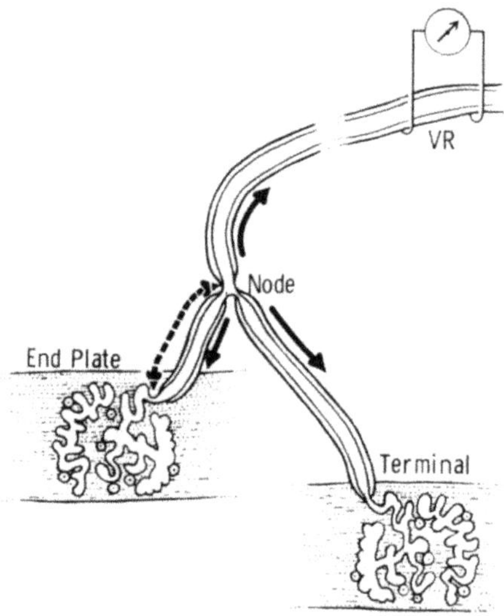

Fig. 2. Scheme for the generation of stimulus-bound repetition by cat soleus motor nerve terminals. After preconditioning, a single stimulus-evoked action potential invades the terminal and causes an interaction between the terminal and the last node of Ranvier. This results in the reexcitation of the axon, and impulses are propagated from that point both ortho- and antidromically. *VR*, ventral root. (Riker WF Jr, Okamoto M 1969)

The repetitive train evoked by high-frequency preconditioning of cat soleus motor nerve terminals typically consists of three to five spikes, ranging in frequency from 200 to 600 Hz (Fig. 1, inset). It can be reevoked with each successive stimulation for some 3–5 min after the preconditioning stimulation. SBR occurrence, therefore, is an in situ signal of a facilitated state of these motor nerve terminals.

3.3 Directly Monitored Motor Nerve Terminals

Soleus (tonic) motor nerve terminals in cats treated with a short intensive triamcinolone regimen (8 mg/kg i.m. daily for 7 days) were found to be much more responsive to frequency facilitation than the untreated normal (Riker et al. 1975; Baker et al. 1978). The data in Table 1 are illustrative. The triamcinolone regimen augmented the excitability of these nerve terminals. There was, for all preconditioning frequencies, a sizeable increase in the number of terminals generating SBR. The downward shift in the frequency facilitation relationship is striking.

An equivalent result was obtained with a single large intravenous dose of methylprednisolone, 90 mg/kg (Table 1) (Baker et al. 1977). This effect was not immediate. It began 8–12 h after steroid administration and peaked at 24 h; its biologic half-life

Table 1. Glucocorticoid effects on cat soleus motor nerve terminal stimulus-bound repetition (SBR).

Drug	No. of axons tested	% axons demonstrating SBR after 10 s conditioning stimulation			
		50 Hz	100 Hz	200 Hz	400 Hz
None	50	12.0	34.0	56.0	80.0
Chronic triamcinolone (8 mg/kg i.m. for 7 days)	112	37.5	68.8	86.6	98.2
24 h after single acute methylprednisolone (90 mg/kg i.v.)	80	35.0	68.8	85.0	100.0
1−60 min after single acute methylprednisolone (90 mg/kg i.v.)	55				30.9

was 48 h. These results disclose that intensive chronic or single dosing with glucocorticoids heightens the excitability of these particular nerve terminals. The PTP findings described on p. 77 and shown in Table 2 reflect this.

Table 2. Single-dose methylprednisolone enhancement of indirectly evoked soleus posttetanic potentiation tested 24 h after sciatic nerve sectioning.

Conditioning frequency (Hz/10 s)	% peak posttetanic potentiation above control value in untreated cats (mean ± SEM) [a]	% peak posttetanic potentiation above control value in methylprednisolone-treated cats (mean ± SEM) [b]
400	110.0 ± 24.2	179.2 ± 26.4
200	69.9 ± 19.5	141.8 ± 15.9 [c]
100	44.2 ± 8.0	118.0 ± 17.1 [c]
50	19.1 ± 2.7	41.9 ± 8.6 [c]

[a] $n = 5$; control is the pretetanic contractile tension.

[b] $n = 6$; methylprednisolone was given intravenously 90 mg/kg immediately after nerve section and, therefore, 24 h prior to testing.

[c] $P < 0.05$, compared with untreated cats.

3.4 Indirectly Monitored Motor Nerve Terminals

The glucocorticoid-enhanced excitability of cat soleus motor nerve terminals was also detected indirectly (Riker et al. 1975). For this purpose, the contractile tension of the soleus muscle was recorded isometrically in response to single supramaximal nerve stimulation (0.5 Hz). In the non-glucocorticoid-treated preparation, a 10-s period of high-frequency (tetanic) stimulation (400 Hz), followed immediately by return to single stimulation, caused an immediate increase in the contractile amplitude, which decayed gradually over 3−5 min. This phenomenon is commonly called posttetanic potentiation (PTP), and occurs in the soleus neuromuscular system as a result of a posttetanic repetitive firing generated by soleus motor nerve terminals, i.e., an SBR. The PTP amplitude, throughout its course, is a sensitive and accurate index of the fraction of a motor nerve terminal population that is generating a posttetanic SBR.

In animals treated with the short intensive triamcinolone regimen, PTP occurred at lower high-frequency preconditionings than in the untreated. Further, after each of these preconditioning frequencies in the treated animals, the PTP amplitudes and durations were significantly greater than in the controls. A typical example of this is shown in Fig. 3. At all but the highest preconditioning frequency (400 Hz for 10 s) the peak PTP was more than triple the control (left column). This reflects an increased number of motor nerve terminals generating SBR and thus their increased excitability in glucocorticoid-treated cats. The result accords with the data obtained with directly sampled axons (Table 1).

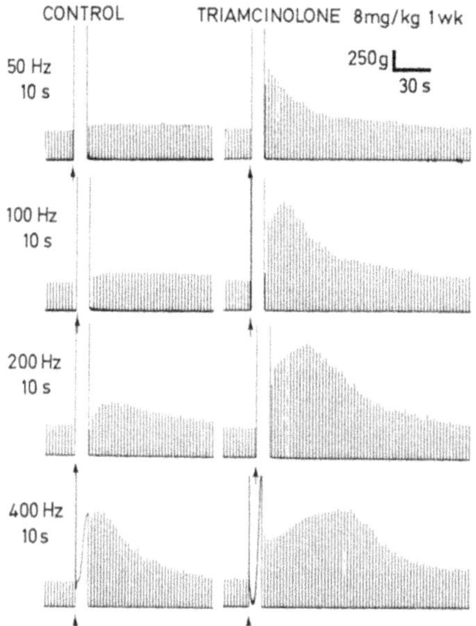

Fig. 3. Triamcinolone augmentation of neurally generated posttetanic potentiation. Soleus motor nerve stimulated with single supramaximal impulses at 0.4 Hz, before and after high-frequency preconditioning. High-frequency stimulation at *arrows;* frequencies listed on *left. Left column,* responses in untreated cat. *Right column,* responses in treated cat, 24 h after cessation of high-dose regimen (8 mg/kg i.m./day for 7 days). (Riker WF Jr, Baker T, Okamoto M 1975)

Just as in the experiments in which motor nerve excitability was monitored via SBR generation, PTP recording also showed that a single large glucocorticoid dose is, with time, sufficient to enhance soleus motor nerve terminal excitability and that this enhancement is dose related (Baker et al. 1977). Thus PTP was increased 24 h after an intravenous dose of 90 mg/kg methylprednisolone hemisuccinate. The PTP augmentations were comparable with those shown in Fig. 3 for triamcinolone. Changes in muscle function per se were ruled out by demonstrating that direct stimulation of the muscle did not augment the PTP following high-frequency conditioning (Baker et al. 1977). Thus the augmented facilitation derives from a neural action of glucocorticoid.

A modification of the technique enabled the neural site of this glucocorticoid action to be narrowed. The sciatic nerve was sectioned and the experiment performed 24 h later. The indirectly evoked contractile response of the cat soleus muscle was negligibly altered at 24 h after nerve section. Relating this circumstance to the fact that a single large intravenous methylprednisolone dose requires 24 h to produce peak enhancement of PTP, it was possible to test the importance of the motor neuron soma in the time-dependent glucocorticoid potentiation of facilitation (Baker et al. 1977). Accordingly, a sciatic nerve was sectioned unilaterally, and the methylprednisolone (90 mg/kg i.v.) given immediately afterwards. After an interval of 24 h, PTP responses were examined bilaterally. From these experiments, it was learned that glucocorticoid augments PTP equally at both the innervated and 24-h denervated junctions (Table 2). It was concluded, therefore, that events in the perikaryon are not involved in the glucocorticoid effect on cat soleus motor nerve excitability and that glucocorticoids act directly on motor nerve axon and terminal.

3.5 Enhancement of Facilitatory Drug Action In Vivo

The effect of glucocorticoids to enhance repetitive generation by cat motor nerve terminals was tested also by a pharmacologic equivalent of frequency facilitation. Drugs such as neostigmine, pyridostigmine, and edrophonium cause mammalian motor nerve terminals (both tonic and phasic) to generate SBR with each single impulse invasion (Riker et al. 1959; Riker 1975). Because of this, these anticholinesterase drugs are also called facilitatory drugs. The neurogenic SBR that follows their administration leads to repetitive transmission and thus to an obligatory potentiation of the muscle response, just as is seen with high-frequency conditioning in the soleus system. As with PTP, the magnitude of the facilitatory drug potentiation reflects the fraction of motor nerve terminals generating SBR. In view of these findings, clinical interactions of glucocorticoids and facilitatory drugs might be expected.

Chronic treatment with triamcinolone (8 mg/kg i.m. daily for 7 days) greatly enhanced the in vivo responsiveness of the cat soleus nerve-muscle system to facilitatory drugs (Riker et al. 1975; Hall et al. 1977a). Figure 4 shows this drug interaction: the edrophonium potentiation of the indirect twitch was nearly triple that of the untreated cat. This indicates that in glucocorticoid-treated animals a greater number of soleus motor nerve terminals respond to edrophonium than in the untreated. Accordingly, the dose-response regression of this edrophonium effect was shifted to the left of that in controls. Edrophonium-evoked SBR recorded from sampled axons was significantly greater in triamcinolone-treated cats, at corresponding doses, than in the untreated cats.

The entire paradigm of this glucocorticoid-facilitatory drug interaction is like that described above for the glucocorticoid enhancement of frequency facilitation (Baker and Riker, unpublished observations). This parallelism was furthered in those experiments in which cats were given the single large methylprednisolone dose (90 mg/kg i.v.) 24 h prior to experiment. Here too, the edrophonium responsiveness of the tonic soleus system was significantly augmented (Baker and Riker, unpublished observations). This was seen both directly and indirectly. For corresponding doses, the incidence of edrophonium-induced SBR in soleus motor axons was higher than in untreated controls and the contractile potentiation greater (Baker and Riker, unpublished observations). Once the triamcinolone or methylprednisolone augmentation of facili-

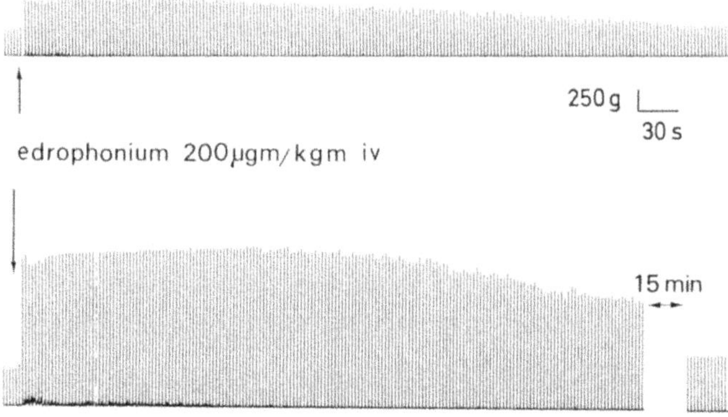

Fig. 4. Effect of triamcinolone on edrophonium potentiation of neurally evoked muscle contractile tension. Soleus motor nerve stimulated with single supramaximal impulses at 0.4 Hz. *Arrows* indicate administration of edrophonium 200 µg/kg i.v. *Upper tracing*, untreated control cat. *Lower tracing*, responses in triamcinolone-treated cat, 24 h after cessation of high-dose regimen (8 mg/kg i.m./day for 7 days). (Riker WF Jr, Baker T, Okamoto M 1975)

tatory drug responsiveness was established in cat soleus motor nerve terminals, it could be maintained with continued glucocorticoid dosing. An antifacilitatory interaction, as described in Sect. 4, did not appear.

Like the facilitatory drugs, guanidine causes cat motor nerve terminals to generate SBR (Feng 1938; Hall 1980a). Also like the facilitatory drugs, guanidine is anti-myasthenic (Minot et al. 1939; Oh and Kim 1973). Further, guanidine increases stimulus-evoked transmitter release (Kamenskaya et al. 1975a, b; Teravainen and Larsen 1975). Recently, Hall (1980a) has shown that an intensive triamcinolone regimen (8 mg/kg i.m. daily for 7 days) significantly enhances guanidine-induced twitch potentiation in cat soleus. This interaction is comparable to that between glucocorticoids and the facilitatory drugs; the contractile potentiation was increased three- to fourfold. Correspondingly, the incidence of SBR in the soleus motor axons was more than doubled.

4 The Antifacilitatory Effect of Glucocorticoids In Vivo

An effect of glucocorticoids to oppose facilitatory drug action was first reported by Patten et al. (1974). In rat neuromuscular preparations in vivo, intravenous cortisol, prednisolone, and methylprednisolone proved equi-effective in acutely suppressing pyridostigmine potentiation of the indirectly evoked twitch. This equivalency contrasts sharply with the fourfold greater glucocorticoid (hormonal) activity of prednisolone and methylprednisolone over cortisol. Further, dexamethasone did not exhibit an antifacilitatory effect. This is entirely disparate with its strong glucocorticoid action. These considerations, together with the speed with which pyridostigmine potentiation is suppressed by the effective glucocorticoids (ca. 1 min) point to a direct depressant action on the neuromuscular apparatus.

Patten et al. (1974) thought that the antifacilitatory effect of these steroids stemmed from a depressant action on motor nerve terminals. They based this partly on the knowledge that facilitatory drugs are known to potentiate the indirectly evoked muscle response by causing mammalian motor nerve terminals to generate SBR (Riker and Okamoto 1969; Riker 1975). A d-tubocurarine-like effect was ruled out because none of these glucocorticoids ever depressed the amplitude of the single isometric muscle response. This accords with the findings of Riker et al. (1975), Baker et al. (1977), Grossie and Albuquerque (1978), and Hall (1980a) with methylprednisolone and triamcinolone, given acutely and chronically, in cats and rats.

The antifacilitatory action of glucocorticoids was confirmed in other studies, which also showed that the facilitatory effects of edrophonium, neostigmine, and pyridostigmine can be acutely erased by certain glucocorticoids (Dengler et al. 1979; Hall 1980b). Prednisolone, given intravenously, reduced or abolished the indirect twitch potentiaton occurring in rat triceps surae after various intravenous neostigmine doses (Dengler et al. 1979). For larger neostigmine doses, greater prednisolone doses were needed for suppression. The prednisolone doses required to suppress neostigmine-induced potentiations ranged from 12.5 to 30 mg/kg i.v. This prednisolone effect developed within 1 min.

The frequency-induced facilitation of cat soleus motor nerve terminals was also antagonized by glucocorticoid (Baker et al. 1977). This was shown both directly by monitoring SBR in nerve terminals (Table 1) and indirectly via PTP recording. Immediately following intravenous methylprednisolone (90 mg/kg), only 31% of soleus motor axons manifested SBR after preconditioning at an optimal frequency; this is to be contrasted with 80% in untreated controls. This methylprednisolone dose also immediately abolished the obligatory PTP in soleus muscle. Because glucocorticoid suppresses frequency-induced facilitation, it appears that its antifacilitatory action occurs at motor nerve terminals. The antifacilitatory interaction of a glucocorticoid with facilitatory drugs has also been demonstrated in vitro (Dengler et al. 1979) (see Sect. 7).

Importantly, Hall (1980b) determined that the twitch potentiation following intravenous edrophonium in cat gastrocnemius-plantaris neuromuscular preparations, a fast (phasic) nerve-muscle system, is reduced in animals pretreated with an intensive triamcinolone regimen (8 mg/kg i.m. daily for 7 days). This contrasts sharply with the previously described facilitatory enhancing effect of this steroid regimen on cat slow (tonic) soleus motor nerves. This difference reemphasizes the point made in Sect. 2: The glucocorticoid action to increase neuronal excitability is not general and excitability can be affected in an opposite fashion in the same or in two functionally related neurons.

5   Time Course of Glucocorticoid Effects on Cat Soleus Motor Nerve Terminals In Vivo

Cat soleus PTP was used to define the time course of the biphasic excitability changes in these motor nerves after high-dose glucocorticoid administration (Baker et al. 1977). The antifacilitatory effect appeared immediately after the intravenous injection of methylprednisolone (90 mg/kg). Recovery from this required 4–8 h, after which the enhancement of the facilitatory responses, as described above, began. This later effect peaked at 24 h and declined to one-half at 48 h. The possible significance of the facilitatory and antifacilitatory actions of glucocorticoids for the treatment of myasthenia gravis is discussed in Sect. 10.

6   Glucocorticoid and *d*-Tubocurarine Antagonisms

d-Tubocurarine (dTC), as a function of dose, competitively and reversibly blocks postjunctional receptors. Although this alone can assure block of neuromuscular transmission, dTC also impairs prejunctional transmitter mechanisms. dTC, as a function of dose and nerve impulse frequency, interferes with transmitter mobilization and transmitter output (Hubbard et al. 1969; Galindo 1971; Blaber 1973; Hubbard and Wilson 1973; Bowman and Webb 1976; Glavinovic 1979; Magelby et al. 1981). Therefore in intact animals or humans, interaction of these pre- and postjunctional dTC effects very likely determines transmission block. In view of these factors and the cited effects of glucocorticoids on motor nerves, it is reasonable to consider that glucocorticoids may influence the transmission-blocking action of dTC as well as its antagonsim.

Arts and Oosterhuis (1975, 1977), using intact mice, demonstrated what appears to be a sharply reduced dTC blocking potency in mice chronically treated with dexamethasone. Their method, however, was indirect. After dTC administration, they measured the time required for control and glucocorticoid-treated mice to regain the ability to walk on a rotating rod. In one set of experiments, prednisolone (ca. 60 mg/kg) was given subcutaneously 3 h prior to dTC (0.4–0.5 mg/kg i.p.). In this circumstance, prednisolone significantly reduced the recovery time from dTC impairment (Arts and Oosterhuis 1975). In a similar experiment, mice were chronically pretreated with dexamethasone (0.15 mg/kg s.c.) three times per week, for periods of up to 50 days. After 30 and 50 days of treatment, average recovery times from dTC impairment were respectively one-third and one-fifth those of the controls (Arts and Oosterhuis 1977).

On the other hand, in rat neuromuscular preparations, the acute administration of glucocorticoids in vivo and in vitro was without effect on dTC block of transmission (Hoffmann 1977). In experiments with phrenic nerve-diaphragm in vitro and rat sciatic nerve–tibialis anterior preparations in vivo, Wolters and Leeuwin (1976) and Leeuwin and Wolters (1977) were unable, with acutely administered glucocorticoids, to affect the magnitude or time course of dTC transmission block. In the in vitro experiment, the diaphragm, prior to dTC exposure, was bathed for 3 min in a Krebs-Ringer solution having a concentration of 2.1 μM prednisolone. In the in vivo experiment, prednisolone and dexamethasone were examined over wide dose ranges. Neither prednisolone (100–3200 μg/kg i.p.) nor dexamethasone (10–320 μg/kg i.p.), injected in conjunction with a full dTC blocking dose, altered the extent or course of the block. Hofmann's (1977) findings with the rat phrenic nerve–diaphragm technique in vitro were similar. Exposure of the preparation to a 0.1 mM prednisolone did not affect the course of dTC block. Further, a continued exposure to prednisolone for 1–3 h did not antagonize the block. Baker and Riker (unpublished observations) found that an intensive triamcinolone regimen (8 mg/kg i.m./7 days) did not affect the dTC LD_{50} in rats.

The experiments of Arts and Oosterhuis (1977), although lacking directness, indicate that mice treated chronically with prednisolone or dexamethasone are strikingly resistant to dTC. The seemingly contradictory results may reflect a species difference between mice and rats or the importance of chronic glucocorticoid dosing.

Recent experiments with cat neuromuscular preparations (Riker et al. 1977; Hall 1980b; Baker and Riker unpublished observations) have demonstrated that chronic intensive glucocorticoid treatment significantly increases the resistance of both tonic and phasic neuromuscular junctions to dTC block. Additionally, these studies showed that the antagonism of dTC by edrophonium is greatly enhanced in the tonic soleus neuromuscular system when the animals are pretreated with glucocorticoid.

In cats chronically treated with triamcinolone (8 mg/kg i.m. daily for 7 days) the in vivo blocking potency of dTC at the tonic soleus junction is significantly reduced. The dose-response regressions for block of indirectly elicited muscle responses in triamcinolone-treated and untreated cats are shown in Fig. 5. It can also be seen that the

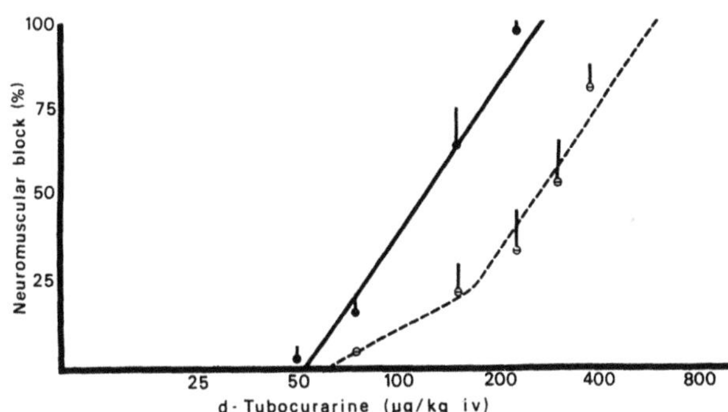

Fig. 5. Triamcinolone effect on the neuromuscular blocking potency of d-tubocurarine. Cat soleus nerve supramaximally stimulated at 0.4 Hz. ●, data (± SEM) from nine untreated controls; ○, data (± SEM) from ten cats 24 h after treatment (8 mg/kg i.m./ day for 7 days)

dose-response relationship for the intravenous dTC blocking action in the triamcinolone-treated animals is biphasic. Important to recognize is that the threshold blocking doses in the treated and untreated do not differ significantly. This suggests, but does not prove, that the chronic triamcinolone treatment has not altered the postjunctional sensitivity to dTC. As dTC dose increases the regression for the treated animals diverges from the control and its slope is flatter. The diverging phase of this regression is a function of dTC dose. Beyond a dTC dose of 225 μg/kg i.v. the regression slope changes and parallels that of the control. Thus the ED_{50} points in the treated and untreated differ by a factor of 2.5. The parallelism of the two regressions suggests that the mechanisms of dTC block are alike in the glucocorticoid-treated and untreated preparations. From these data, it is concluded that the dTC blocking potency is reduced in the soleus nerve–muscle system of triamcinolone-treated cats.

The initial flatter-sloped regression, which develops as a function of dTC dose in the triamcinolone-treated cats, suggests that dTC is in some manner antagonizing its own postjunctional blocking action. Considering that the monoquaternary ammonium portion of the dTC molecule embodies a phenolic quaternary ammonium ion resembling edrophonium, it seemed reasonable to us that in glucocorticoid-treated cats a latent facilitatory action of dTC may have been exposed. This would be consistent with the documented exaggeration of facilitatory agencies by glucocorticoids at this tonic junction. Accordingly, in triamcinolone-treated cats, dTC-evoked SBR could always be found in some soleus motor axons; its incidence in sampled axons was 19% following dTC doses of 50–225 μg/kg i.v. Thus, dTC often caused potentiation of the indirectly evoked contractile response in these glucocorticoid-treated animals. This dTC-evoked potentiation preceded or followed a dTC blocking action, if the latter appeared; often potentiation developed without block. When block occurred, the time for recovery was significantly shortened. It is concluded, therefore, that dTC can act as a partial facilitatory agonist and that this property visibly emerges in the cat tonic soleus system exposed to chronic triamcinolone treatment.

Hall (1980b) reported a similar glucocorticoid–dTC interaction in the phasic neuromuscular system of cats treated with triamcinolone (8 mg/kg i.m. daily for 7 days). The intravenous dTC blocking potency was halved and the duration of block significantly decreased. The dose-response regressions (Hall 1980b; Fig. 2) are essentially like those shown for the soleus neuromuscular preparation in Fig. 5. Although Hall (1980b) saw in the triamcinolone-treated cat an occasional slight twitch potentiation with threshold blocking doses of dTC, he did not search for SBR. He concludes firstly that dTC acts as a partial facilitatory agonist on cat phasic motor nerve terminals, as reported for the tonic (Riker et al. 1977). Secondly, he believes that glucocorticoid pretreatment, by enhancing responsiveness to the prejunctional facilitatory action of dTC, increases transmitter output and thereby opposes the postjunctional dTC block.

In unpublished experiments in our laboratory, the anticurare action of edrophonium, in the soleus neuromuscular preparation, was seen to be greatly enhanced by pretreating cats intensively with triamcinolone (8 mg/kg i.m. daily for 7 days). In treated animals, dTC transmission blockages of 70%–80% were antagonized completely within 10 s of an intravenous dose of 25 μg/kg edrophonium. The antagonism was permanent. In non-glucocorticoid-treated cats full restoration of this degree of dTC block required a 200–400 μg/kg dose of edrophonium. Clearly, the positive glucocorticoid-facilitatory drug interaction, as seen in the absence of dTC, prevails in the presence of dTC. This interaction remains to be tested in a fast nerve-muscle system.

The increased anticurare effectiveness of edrophonium is thought to reflect a heightened excitability of motor nerve terminals to edrophonium, which in turn leads to an increased transmitter output. A postjunctional change in receptor affinity for dTC, consequent to the glucocorticoid treatment, is a possible alternative. However, in vitro studies (vide infra) do not favor this explanation.

7 Glucocorticoid Effects on Neuromuscular Function In Vitro

A number of studies, using nerve-muscle preparations in vitro, reveal a complexity of glucocorticoid effects when these steroids are added acutely. The in vitro experiments in which glucocorticoids were added directly to the bathing fluid lend support to the notion that nerve membranes can be directly affected by these steroids. In these experiments, the nerve terminals were disconnected from their cell bodies prior to the initial exposure to the steroid. In vitro preparations permit accurate control of steroid concentrations, the use of intracellular measurements from the muscle membrane, and the recording of nerve-evoked endplate currents under voltage-clamp conditions. Experiments utilizing this technique first disclosed that glucocorticoids may directly affect transmission, and motor nerve terminals in particular (Wilson et al. 1974). These experiments have since been confirmed and extended.

Miniature endplate potential (mepp) frequency is an index of motor nerve terminal function. An increase in mepp frequency has been observed following the acute in vitro addition of glucocorticoids to various nerve—muscle preparations: the rat diaphragm (Wilson et al. 1974; Dengler et al. 1979; van Wilgenburg 1980; Dlouha and Vyskocil 1979), mouse diaphragm (Dudel et al. 1979), rat anterior tibialis (Leeuwin et al. 1978), and rat flexor digitorum longus (Kim et al. 1979). Increased mepp frequency has also been reported in the rat diaphragm after intensive in vivo treatment with prednisolone (Hoffman 1977) and in the rat extensor digitorum longus after 4 days of triamcinolone dosing (Grossie and Albuquerque 1978). The mepp frequency increases which follow either acute or chronic glucocorticoid administration signal an action on motor nerve terminals.

The effect of glucocorticoid to antagonize facilitatory drug action on motor nerve terminals has also been demonstrated with rat diaphragm in vitro (Dengler et al. 1979). A neostigmine-induced SBR in phrenic nerve was employed to generate an increased contractile tension known to arise from a motor nerve terminal facilitation. The addition of prednisolone depressed the contractile force reflecting its acute antifacilitatory action. However, the contractile force was decreased disproportionately more than the SBR. This led to the thought of an additional postjunctional blocking action of glucocorticoid. However, this complex effect could also reflect interactions between the prejunctional facilitatory action of neostigmine, its acute antagonism at that site by prednisolone, and the direct neuromuscular blocking action of neostigmine.

Both facilitatory and antifacilitatory actions of prednisolone, and their dependence on glucocorticoid concentration (dose), were revealed in a study by Dreyer et al. (1979) on the frog cutaneous pectoris neuromuscular junction. At low concentrations of prednisolone (1 μM), the nerve-evoked peak endplate currents recorded in a voltage-clamped muscle were increased two-fold. This reveals a direct acute facilitating effect of prednisolone on neuromuscular transmission. Furthermore, the sensitivity of the postjunctional membrane to iontophoretically applied acetylcholine (ACh) remained unchanged. Consequently, it became clear that prednisolone, at low concentrations, increased the nerve-evoked release of transmitter.

At higher concentrations of prednisolone (40 and 100 μM) postjunctional effects appeared (Dreyer et al. 1979). The peak endplate current evoked by quantitative iontophoresis of ACh fell to about 50% of the control value. Surprisingly, the time-to-peak endplate current was shortened at the higher ACh concentrations. From these findings, it is evident that an acute direct effect of glucocorticoid on neuromuscular transmission might be complicated by a combination of pre- and postjunctional actions, depending on the concentration of the steroid. The motor nerve terminals appear to be the more susceptible junctional element.

An acute glucocorticoid-induced depression of postjunctional sensitivity, especially with bath-applied concentrations higher than 64 μM, has been observed in the form of reduced mepp amplitude and reduced epp size. The epp quantal content was unchanged (Wilson et al. 1974; Kim et al. 1979; Dudel et al. 1979; van Wilgenburg 1979, 1980; Dlouha and Vyskocil 1979). This is in contrast to the results of Dreyer et al. (1979), and suggests that the stimulus-evoked output of transmitter is not always increased by glucocorticoids.

The observation by Dudel et al. (1979) that prednisolone (1 mM) reduces the decay time constant of the mepp (a measure related to the mean open time of the ACh-activated ion channel) offers a possible explanation for the decreased postjunctional sensitivity. This observation also suggests that the kinetics of the ACh-activated ion channels are altered by glucocorticoids. Consistent with this suggestion is the significant reduction in the epp rise time to peak, as noted by Kim et al. (1979) in the rat, and the reduction in time-to-peak of the endplate current, as observed by Dreyer et al. (1979) in voltage-clamped frog muscle.

Changes in the passive properties of the muscle membrane or in its action-potential generating mechanism do not account for the reduced response of glucocorticoid-treated muscles to ACh (Dlouha and Vyskocil 1979; Grossie and Albuquerque 1978; Dengler et al. 1979). Glucocorticoid effects on the resting potential of muscle fibers depend on the steroid concentration. Low concentrations of cortisol (2 μM) produce a 3–4 mV hyperpolarization. Higher concentrations (20 μM) produce either no effect or a small depolarization (10 mM) (Dlouha and Vyskocil 1979). At a concentration of 0.5–1.0 mM prednisolone there is no effect on the muscle fiber resting potential (Wilson et al. 1974; Kim et al. 1979). The mean resting potential of muscle fibers was decreased from −78 to −71 mV during an intensive 4-day treatment with triamcinolone in the rat (Grossie and Albuquerque 1978). Although the ionic basis of these changes in membrane potential remains to be elucidated, in vitro studies do not indicate that glucocorticoids exert major effects on postjunctional membrane.

8 Glucocorticoids and Choline Uptake

The results on neuromuscular transmission described above raise the possibility of a glucocorticoid effect on the synthesis or release of ACh. Based on the premise that the rate-limiting step in the synthesis of ACh is the recapture of choline from the intracellular cleft by a high-affinity choline uptake system, some investigators have studied the modification by glucocorticoids of the actions of hemicholinium-3 (HC-3), a potent inhibitor of the choline uptake system (Arts and Oosterhuis 1975; Wolters and Leeuwin 1975, 1976; Veldsema-Currie et al. 1976; Leeuwin and Wolters 1977; Leeuwin et al. 1978). Using the rat diaphragm and the anterior tibialis muscle in vitro, these investigators found that dexamethasone and prednisolone lessened or reversed the neuromuscular block due to HC-3. Further, glucocorticoids raised the LD_{50} of HC-3 in rats and mice (Arts and Oosterhuis 1975; Wolters and Leeuwin 1975). They suggested that glucocorticoids reduced the affinity of HC-3 for the choline carrier (Veldsema-Currie et al. 1976; see Hoffman 1977 for a failure to reproduce some of these findings).

In view of these findings and the unusual effect of glucocorticoids to enhance frequency and drug-induced facilitations in cat soleus motor nerves Riker et al. (1979) examined the influence of chronic glucocorticoid dosing on high-affinity choline accumulation (HACA) in synaptosomal preparations of cat brain. After treating cats daily for 1 week with either triamcinolone (8 mg/kg) or cortisol (4 or 32 mg/kg), HACA was elevated by 37%–75% in synaptosomes derived from the caudate-putamen. The changes in the caudate-putamen uptake were attributable to a 50% increase in the maximal transport velocity with no significant change in the apparent transport constant for choline. This work also disclosed pronounced regional differences, since significant increases in HACA were not seen in the hippocampus-fornix, prefrontal cortex, or the anterior perforated space. Both the IC_{50} and the Ki for HC-3 were unchanged in caudate-putamen and hippocampus-fornix synaptosomes after treatment of the animal with cortisol (32 mg/kg for 7 days).

Acute treatment of cats with a single intravenous dose of methylprednisolone (90 mg/kg) produced, after 3 h, 75%–84% increases in HACA in synaptosomes from caudate-putamen and hippocampus-fornix. At 24 h after treatment, HACA was increased in synaptosomes prepared from caudate and the anterior perforated space. HACA in the hippocampus-fornix synaptosomes had returned to baseline values. Thus the glucocorticoid-induced increase in HACA in caudate-putamen synaptosomes is sustained, while this effect is transient in the other brain regions examined. It is noteworthy that none of these in vivo glucocorticoid effects could be produced by addition of the steroids in vitro or by alterations of the Na^+ and K^+ levels in the test tubes. These experiments underscore the differences of cholinergic neurons in different parts of the brain, in analogy with the observed differences between the cholinergic motorneurons that innervate phasic and tonic muscles in the cat. The time course of the HACA experiments does not permit any conclusions to be drawn as to the cellular site or mechanism of action. However, glucocorticoid receptors have recently been identified in the caudate (Turner and DeFiore 1980). Nevertheless, changes in the affinity of choline carrier, as measured by HC-3, can be excluded as a possible mechanism of glucocorticoid action.

9 Glucocorticoid Effects on Spinal Segmental Reflexes

Both acute and chronic glucocorticoid treatment affect spinal cord segmental reflex functions in the cat. The spinal cord segmental reflex constitutes a reflex arc which in its simplest form involves only one synapse and two neurons: a monosynaptic reflex pathway (MSR). When a greater number of synapses and neurons are involved, the reflex is said to be polysynaptic. Hall and Baker (1979a) found that intravenous methylprednisolone (30, 60, and 90 mg/kg) promptly increases the amplitude of the MSR discharge evoked by a stimulus applied to the afferent nerves. This glucocorticoid effect was dose dependent and the augmented output lasted for 10–60 min. Polysynaptic discharge was also increased by these doses but to a lesser extent and for a shorter period than the MSR output (Hall and Baker 1979a). Short intensive glucocorticoid treatment (triamcinolone 8 mg/kg for 7 days) also increased both the MSR and polysynaptic discharges in response to spaced single stimuli (Hall et al. 1978).

Following high-frequency conditioning (50–800 Hz/10 s) of the primary (Ia) afferent endings, a potentiation of the MSR response (PTP) occurs in the cat. The MSR PTP is directly proportional to the conditioning frequency (Lloyd 1949). The triamcinolone treatment also enhanced the MSR peak PTP amplitudes at all test frequencies (Fig. 6). This is similar to the effect of this glucocorticoid treatment on

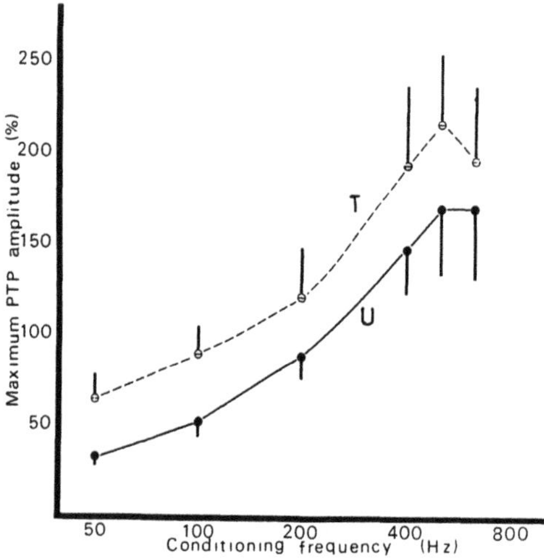

Fig. 6. Triamcinolone effect on monosynaptic reflex posttetanic potentiation (PTP). Peak PTP amplitude generated after each 10-s conditioning frequency is plotted as a percentage above the pretetanic control amplitude (± SEM). Data from 18 untreated controls *(U)* and 12 cats 24 h after treatment *(T)* (8 mg/kg i.m./day for 7 days). Hall ED, Baker T, Riker WF Jr 1978)

the SBR and PTP responses of the cat soleus neuromuscular system (Riker et al. 1975). Specifically, this glucocorticoid treatment shifted the frequency-response curve for MSR PTP downward, as it did for the cat tonic soleus neuromuscular system. Concurrently, the polysynaptic discharge was also enhanced. The durations of both the MSR and polysynaptic PTP were greatly prolonged. Since the augmented MSR output in the glucocorticoid-treated cats results from an increased number of units actively transmitting, the enhanced MSR PTP was concluded to represent a heightened excitability of the Ia afferent endings. This would also account for PTP prolongation.

Neither postsynaptic nor recurrent inhibition was affected by glucocorticoid treatment. Presynaptic inhibition, on the other hand, was greatly increased. At shorter time intervals between stimulation of antagonist and agonist afferent nerves, presynaptic inhibition was twice that seen in untreated controls. The duration of the presynaptic inhibition in glucocorticoid-treated cats, however, did not differ from that of the untreated control (Hall et al. 1978). The fast decay of presynaptic inhibition in glucocorticoid-treated animals reflects, in another way, the predominance of increased excitability of the Ia afferent endings.

An increase in the dorsal root depolarization potentials (phase V, as described by Lloyd and McIntyre 1949) occurred in triamcinolone-treated cats and this correlates with the increased polysynaptic inhibition. Since the dorsal root potential represents the threshold depolarization of unstimulated Ia endings by activity evoked in adjacent endings, the glucocorticoid-induced increase in the excitability of the Ia endings was disclosed in yet a third way.

The fraction of the total population of Ia afferent endings stimulated by increasing current strengths and the resultant fraction of the maximal MSR discharge describes the MSR input—output relationship. The triamcinolone treatment caused this input—output relationship to shift such that less Ia activation was needed to initiate a threshold MSR discharge (Hall and Baker 1979b).

Repeated stimulation of Ia afferent endings causes a gradual rundown and eventual leveling of the MSR discharge. Presumably, this reflects transmitter depletion. Triamcinolone treatment prevented rundown of the MSR discharge. Furthermore, as the frequency of stimulation was increased, less MSR rundown occurred in the glucocorticoid-treated animals than in untreated controls (Hall and Baker 1979b).

Analysis of the apparent transmitter turnover parameters revealed that the rate of replenishment of the store of transmitter immediately available for release was more than doubled by triamcinolone treatment, while the fractional release of transmitter per impulse by the spinal Ia afferent terminals was not affected (Hall and Baker 1979b). The decrease in Ia activation necessary to evoke a threshold MSR response, without an increase in the fractional release of transmitter, represents an improvement in the safety factor for terminal impulse invasion. The improved maintenance of repetitive transmission reflects an increased rate of transmitter mobilization. These workers concluded that this glucocorticoid enhancement of synaptic transmission in the spinal cord may be due, in large part, to improved transmitter turnover.

The parallelism between glucocorticoid effects on cat spinal Ia afferents and tonic soleus motor nerves supports the earlier suggestion that soleus motor nerve terminal responsiveness to glucocorticoids may be representative of that of certain CNS nerve endings. On the other hand, the unresponsiveness to glucocorticoids of nerve endings mediating postsynaptic and recurrent inhibition reemphasizes that only certain nerve endings in the CNS are susceptible to glucocorticoids, comparable to the observed differences in reactivity between tonic and phasic motor nerves.

10 Considerations of the Neuromuscular Effects of Glucocorticoids for Their Use in Myasthenia Gravis

10.1 Myasthenic Exacerbation

Since an impressive array of evidence has established the autoimmune nature of myasthenia gravis, it follows that those clinical studies that have proved the benefits of glucocorticoids in the treatment of this disease also logically conclude that their therapeutic effectiveness results mainly from their immunosuppressant actions. However, the presumably direct actions of glucocorticoids on neuromuscular function discussed in this chapter suggest that these too may contribute to their effectiveness in the treatment of myasthenia gravis.

Current treatment of myasthenia gravis with glucocorticoids has evolved from clinical investigations conducted more than 30 years ago, in which corticotropin dosing was found to alleviate the disease (Simon 1935; Torda and Wolff 1949, 1951; Schlezinger 1952). Shortly thereafter, however, other studies disclosed that this therapy often exaggerated myasthenic weakness (Shy et al. 1950; Millikan and Eaton 1951; Grob and Harvey 1952). On occasion serious respiratory distress and death occurred. These adversities slowed exploration of corticotropin's possible benefits until the 1960s when certain investigators reported that short intensive courses of corticotropin are therapeutically effective (Freydberg 1960; von Reis et al. 1965; Osserman and Genkins 1966; Grob and Namba 1966; Namba et al. 1971).

In 1971, Namba et al. reviewed the benefits as well as the limitations consequent to 274 courses of corticotropin in myasthenia gravis. Apart from the clear benefit of intensive corticotropin therapy, they reported that the most common untoward effect was a transient decrease in muscle strength. This was reported in 72%; in their own series, this adversity appeared in 96% of the initial treatments, 45% being severe. In patients with severe myasthenia, a nearly complete paralysis might occur in conjunction with a loss of responsiveness to anticholinesterase (facilitatory) drugs, e.g., neostigmine. The decrease in muscle strength emerged during the first 5 days of corticotropin treatment and peaked between the 2nd and 10th days.

The corticotropin studies are cited because it is generally recognized that the antimyasthenic effectiveness of this trophic hormone is secondary to the release of endogenous cortisol (Namba et al. 1971). Thus, the initial transient impairment of neuromuscular transmission caused by corticotropin appears attributable to cortisol (Grob and Namba 1966).

Subsequent clinical experiences with the glucocorticoids (prednisolone, methylprednisolone, and dexamethasone) have not essentially differed from the corticotropin findings (Jenkins 1972; Brunner et al. 1976; Mann et al. 1976; Johns 1977). The incidence of increased weakness during high-dose therapy with these glucocorticoids, in patients with mild to severe generalized myasthenic disease, reportedly ranges from 44% to 82%. Occasionally, severe weakness caused by the glucocorticoids necessitated respiratory support, and extensive neuromuscular paralysis has occurred. Because of this adversity, all recommend that glucocorticoid therapy of the myasthenic patient be initiated in the hospital.

Glucocorticoid exacerbation of myasthenia has been seen as early as 8 h after dosing (Jenkins 1972) but is most common between the 3rd and 5th days (Brunner et al. 1976; Mann et al. 1976; Johns 1977). However, it may appear as late as the 10th to 21st days. The duration of the exacerbation, as given by Mann et al. (1976) and Johns (1977) for a series of 47 patients receiving 67 courses of prednisone treatment, ranges from 60 min to 20 days.

Early exacerbation of weakness consequent to corticotropin or glucocorticoid administration was demonstrable via electromyography, as a transmission deficit. Thus a sharp decrement in indirectly evoked muscle action potentials was seen during the period of steroid-induced weakness (Gandiglio and Pinelli 1968). Such a decrement is characteristic of myasthenia gravis, but was especially exaggerated at the time of the crisis.

Clinical reports indicate that the occurrence of myasthenic exacerbation is related to glucocorticoid dose (Kjaer 1971; Seybold and Drachman 1974; Brunner et al. 1976; Mann et al. 1976; Johns 1977). Accordingly, myasthenic exacerbation appears less frequently with alternate-day prednisone, especially if the dose is on the low side (Seybold and Drachmann 1974; Brunner et al. 1976). Beginning with low-dose alternate-day prednisone (15 mg), Brunner et al. (1972, 1976) gradually increased the dose in three patients with mild to severe myasthenia, encountering exacerbation when the dose reached 25–37.5 mg. Considering then that the glucocorticoid exacerbation of myasthenia is dose dependent, its less than 100% occurrence can be attributed to: (a) differing glucocorticoid doses in the several studies, (b) individual variation in exacerbation of myasthenic weakness, (c) severity of the disease, and (d) presence or absence of concurrent facilitatory drug treatment. These same factors probably also account for the varying times at which the exacerbation appears and peaks.

Patten et al. (1974) first proposed that the glucocorticoid-induced myasthenic exacerbations may result from an acute glucocorticoid block of antimyasthenic drug (i.e., facilitatory drug) effects. His proposal was based on those cited experiments showing this antagonism in rats. His findings have since been confirmed in cat neuromuscular preparations (Baker et al. 1977) and in an in vitro rat nerve-muscle preparation (Dengler et al. 1979). These latter reports are especially notable because they demonstrate that glucocorticoids act on motor nerve to antagonize both facilitatory drug action and manifestation of a neuronally generated facilitation (Baker et al. 1977; Dengler et al. 1979). These experiments also establish, in accord with the clinical reports, that the antifacilitatory action of glucocorticoid is dose related (Patten et al. 1974; Baker et al. 1977; Baker and Riker, unpublished observations).

The antifacilitatory effect of glucocorticoids is seen within 1 min in experimental animals (Patten et al. 1974; Baker et al. 1977). A reasonable clinical approximation of the experimental demonstration of antifacilitation is Jenkins' (1972) report of a severe exacerbation 8 h after an initial oral prednisone dose (100 mg) in a patient receiving facilitatory drug medication. It is important to note, therefore, that in almost all of the reported clinical trials with glucocorticoids facilitatory drug medication was concurrently administered. Accordingly, it has been established that during the period of clinical exacerbation the responsiveness to facilitatory drug treatment is either diminished or lost. Thus Jenkins (1972) and Brunner et al. (1976) demonstrated a complete lack of responsiveness to intravenous edrophonium during severe glucocorticoid-induced exacerbations of myasthenia. This antifacilitatory action of glucocorticoids is evidently direct and probably occurs at motor nerve endings in view of the experimental demonstrations cited in the previous section.

Although there seems little doubt that glucocorticoid antagonism of facilitatory drug action can precipitate myasthenic weakness, this exacerbation can also occur in patients not treated with facilitatory drugs. Brunner et al. (1976) treated nine patients suffering from mild to severe generalized myasthenia gravis solely with prednisone (100 mg), daily or on alternate days. Severe initial weakness occurred in two of four patients dosed daily, and all five patients receiving alternate-day dosing became slightly weaker. A larger clinical experience with glucocorticoid dosing initiated in the absence of facilitatory drug treatment is needed to evaluate the role of a depressant action of glucocorticoids in the exacerbation syndrome. Nevertheless, since facilitatory drug was not present in these seven patients, it is suggested that the glucocorticoid exacerbation of myasthenia represents a dose-dependent suppression of an endogenous facilitatory mechanism. The acute suppresion of frequency facilitation by methylprednisolone or cortisol may reasonably exemplify a glucocorticoid antagonism of an intrinsic facilitatory mechansim (Baker et al. 1977).

The antifacilitatory effects of the glucocorticoids as determined in experimental animals (Patten et al. 1974; Baker et al. 1977; Dengler et al. 1979; Hall 1980b) appear to correspond closely, if not directly, with the clinical adversity of glucocorticoid exacerbation of myasthenia.

10.2 Glucocorticoid Improvement of Neuromuscular Function

In the extensive and representative studies of Brunner et al. (1976), Mann et al. (1976), and Johns (1977), it is evident that high-dose glucocorticoids eventually increase muscle strength and cause disease remission in 86%–100% of treated patients. This extent of improvement was seen with intramuscular methylprednisolone, oral dexamethasone, oral prednisone, and intravenous or intramuscular corticotropin. These studies also establish the important fact that the beneficial effect of the glucocorticoids is dose and time related. It now seems generally agreed that all patients with myasthenia gravis, except for those with ocular involvement only, are best started on daily high-dose prednisone (60 mg) until there is a sustained improvement (Brunner et al. 1976; Mann et al. 1976; Johns 1977). Dosing is subsequently tapered to a maintenance level given on alternate days.

Clinical improvement in muscle strength, evolving with either corticotropin or prednisone therapy, is also demonstrable electromyographically. The characteristic rundown of indirectly and repetitively evoked muscle action potentials seen prior to therapy is largely repaired at times of peak improvement (Namba et al. 1967; Namba 1972; Gandiglio and Pinelli 1968; Pinelli et al. 1974).

The effectiveness of low and high doses of prednisone and dexamethasone was specifically examined in a small number of patients by Brunner et al. (1976). With prednisone doses as low as 15 mg daily and small gradual dose increments, improvement was slower and less marked than in those treated with larger doses. Johns (1977) noted that reduction of prednisone dose also soon led to bouts of weakness.

The importance of dose is emphasized further by experiences with alternate-day dosing. On this schedule, Seybold and Drachman (1974), using varying low doses of prednisone (25 mg, increasing gradually to 100 mg, on alternate days), found that myasthenic weakness usually increased on the "off" day. The "on–off" phenomenon with alternate day dosing has been seen by others (Jenkins 1972; Brunner et al. 1976; Mann et al. 1976; Howard et al. 1976; Johns 1977). The intermittent weakness was offset when low supplements of prednisone were given on the off day. Clearly, improvement in strength during the course of therapy is closely tied to dosing interval.

Although the magnitude of glucocorticoid dose is critical in overcoming myasthenia weakness, variable periods of time elapse before improvement is first detected. Although initial improvement is occasionally seen as early as 3 days, the mean onset time is 13 days (range: 3–30 days) (Jenkins 1972; Pinelli et al. 1974; Brunner et al. 1976;

Mann et al. 1976; Johns 1977). Maximum improvement with prednisone (10–100 mg p.o. daily and on alternate days, average 60 mg) occurred between 3 and 90 days after beginning dosing (Brunner et al. 1976; Mann et al. 1976; Johns 1977).

The link between dose and time is evident from the results of alternate-day treatment. In a comparative study, Brunner et al. (1976) determined that the daily administration of a glucocorticoid led to a more rapid improvement than the alternate-day regimen. Maximum improvement was achieved in 92% of the effective courses of daily glucocorticoids within 15–20 days; in contrast only 34% on alternate-day regimen reached maximum improvement in this period.

The dose and time requirements for a maximal therapeutic effect present two important considerations: (1) a critical level of steroid cumulation and (2) a change in motor nerve function with time. These pharmacologic and functional concepts would relate to the need for maintenance glucocorticoid dosing as long as the disease is active.

10.3 Comparison of Clinical and Animal Studies

The studies of glucocorticoid effects on neuromuscular function in normal animals provide certain correlates to the clinical experiences (see Sect. 6). Especially relevant are those experiments demonstrating that chronic treatment with glucocorticoids decreases with time the paralytic effect of dTC. Functionally, dTC remarkably mimics myasthenia gravis. In cats, motor nerve terminals appear to be the site of the triamcinolone antagonism of dTC (Riker et al. 1977; Hall 1980b; Baker and Riker, unpublished observations). That motor nerve terminals are involved in this action is suggested by the uncovering of a partial facilitatory antagonism of dTC in glucocorticoid-treated cats (Riker et al. 1977).

In the cited clinical studies, improvement of myasthenic patients was uniformly associated with a decreased requirement for facilitatory drug medication. This was usually striking and in many patients facilitatory drug medication was eventually discontinued. When it was continued, the required dosage was sharply reduced. Several patients treated with prednisone (100 mg on alternate days) for long periods responded to pyridostigmine at doses as low as 15 and 30 mg (Engel 1976). With a similar regimen, Jenkins (1972) was able to halve the facilitatory drug dosage. The above investigators state that many patients on prednisone develop an increased responsiveness to facilitatory drugs before muscle strength improves.

An interesting aspect of this positive drug interaction is the opportunity to improve and maintain the antimyasthenic effectiveness of facilitatory drug via modest chronic glucocorticoid dosing. Kjaer (1971) treated three patients who were responding poorly or not at all to facilitatory drugs. He administered prednisone (45 mg) daily, while maintaining their facilitatory drug dosage. After 1 week, he progressively tapered the prednisone to 10–15 mg daily. During this period the patients developed full responsiveness to the facilitatory drug with complete symptomatic relief. He further demonstrated the interactive role of prednisone by discontinuing it. With this change, both responsiveness to facilitatory drug and control of the myasthenia were lost. Although this is a small number of patients, Kjaer's (1971) results show the opportunity for a treatment regimen based on the positive glucocorticoid–facilitatory drug interaction.

An important advantage may be the avoidance of undesired effects of steroid or facilitatory drugs.

The increased reponsiveness of the myasthenic patient to facilitatory drugs during chronic glucocorticoid treatment is comparable to the similar positive interaction that occurs in the normal cat (Riker et al. 1975, 1977; Baker et al. 1977, 1978; Hall et al. 1977a, b). Under both laboratory and clinical conditions, the increased responsiveness to facilitatory drug develops as a function of glucocorticoid dose and time. In the cat, the glucocorticoid enhancement of facilitatory drug action develops over a period of time and is always preceded by an antifacilitatory drug action, as previously described (Baker et al. 1977). The shortest time to the peak of facilitatory enhancement, as measured in the tonic soleus system, proved to be 24 h after a large intravenous dose of methylprednisolone (90 mg/kg) (Baker et al. 1977). With a smaller methylprednisolone dose (32 mg/kg) given intramuscularly, a comparable peak was reached after seven daily doses. For triamcinolone given intramuscularly, equivalent peaks of enhanced facilitation were reached in 7 and 14 days after doses of 8 and 4 mg/kg, respectively; a dose of 8 mg/kg i.m. for 3 days gave one-half the peak level of glucocorticoid enhancement. Thus the glucocorticoid enhancement of neuromuscular facilitation in the cat is dose and time dependent, as is the enhanced responsiveness to facilitatory drugs in the steroid treatment of myasthenia gravis.

The positive interaction between chronic glucocorticoid dosing and facilitatory drugs is also evidenced by a remarkable prolongation of edrophonium action. In three myasthenic patients receiving chronic high-dose prednisone, intravenous edrophonium in doses as low as 2 mg improved and maintained vital capacity and the transmitted response to repetitive nerve stimulation for as long as 2 h (Rosenbaum et al. 1975). This clinical result correlates directly with the effect of chronic glucocorticoid dosing to augment and prolong intravenous edrophonium action on cat soleus motor nerve terminals (Riker et al. 1975; Hall et al. 1977a). In view of these experimental and clinical findings, motor nerve terminals should not be overlooked as a site of both glucocorticoid and facilitatory drug actions.

A glucocorticoid-induced increase in the excitability of motor nerve terminals in myasthenic patients is evident in the reports of Brunner et al. (1976), Engel (1976), and Drachman (1978). They note that facilitatory drugs cause fasciculations in myasthenic patients chronically dosed with prednisone. This is an unusual response, because facilitatory drugs rarely evoke fasciculations in myasthenic patients (Gammon and Scheie 1937; Harvey and Lilienthal 1941; Harvey et al. 1941; Osserman and Kaplan 1952). Facilitatory-drug-evoked fasciculations are generated by motor nerve terminals (Masland and Wigton 1940; Riker 1966). Both fasciculations and the SBR phenomenon are manifestations of the same prejunctional function. Drug-evoked fasciculations represent random, spontaneous, asynchronous repetitive discharges of motor nerve terminals. Facilitatory-drug-evoked SBR is perforce a synchronized fasciculaton; it is a synchronous repetitive afterdischarge in response to a simultaneous excitation of all motor nerves. From this basic information and the clinical data, it appears that glucocorticoids increase the excitability of motor nerve terminals in the myasthenic patient. Those glucocorticoid effects in experimental animals that resemble glucocorticoid effects in myasthenia gravis are listed in Table 3.

Table 3. Comparable glucocorticoid effects on neuromuscular function in experimental animals and myasthenia patients.

Glucocorticoids in experimental animals	Glucocorticoids in patients with myasthenia gravis
Prompt depression of motor nerve terminal excitability, reflected by a decrease in mepp frequency and SBR suppression	Early exacerbation of myasthenia weakness
Blocking action of dTC decreased by chronic treatment (mice and cats)	Transmission improved by chronic treatment
Facilitatory drug (anticholinesterases) action on motor nerve terminals potentiated in effect and time	Decreased requirement for facilitatory drug (anticholinesterases) therapy
Potentiation dependent on dose and duration of glucocorticoid treatment	Effect dependent on dose and duration of glucocorticoid treatment

11 Speculations

The reader is reminded of the prefatory remarks: Direct steroid actions on membranes are not considered as alternatives to the well-documented effects mediated via cytoplasmic receptors and nuclear actions. Rather, direct effects are regarded as additional to classic mechanisms. This particular view immediately raises the pharmacologic issue of specificity. In this context, a specific drug action usually connotes interaction with a definite biochemical entity, e.g., a receptor or an enzyme. Nonspecific drug action, on the other hand, is usually thought of as resulting from physical effects of the drug, e.g., general anesthetics. Neural effects of glucocorticoids and other steroid hormones which are receptor mediated, i.e., specific in action, have been described and discussed extensively by McEwen et al. (1979).

It has been suggested (Towle and Sze 1978; McEwen 1981) that some direct effects of steroids on excitable membranes may be mediated via a still uncharacterized membrane-bound receptor. There is yet little direct data to support such a proposal; definitive experiments are difficult due to the lipophilic nature of the available ligands.

If the characteristic neuropharmacologic effects of glucocorticoids, as reviewed here, were to arise from interaction with a specific membrane-bound receptor, this protein would have to exhibit and effect the following: (a) a much lower affinity for glucocorticoids than the cytoplasmic receptor; (b) an equal affinity for the natural and synthetic glucocorticoids; (c) a lack of structural specificity for the observed anti-facilitatory effect with respect to the 11α- and 11β-cortisol epimers (Baker and Riker, unpublished); (d) interaction with nonhormonal anesthetic steroids; and (e) induce-sequential biphasic effects in axons separated from their somata. This would be an unusual set of properties in view of what is presently known about steroid-binding proteins.

Enhancement of neuronal excitability by glucocorticoids thus far has been most clearly defined using the SBR generating property of cat motor nerve terminals. This enhancement consists of an increased number of frequency-conditioned terminals generating SBR (see Table 1). To appreciate this, it should be recalled that excitation of all axons in the motor nerve initiates action potentials that invade all their terminals. However, only 34% of a motor nerve terminal population, preconditioned with stimuli at 100 Hz for 10 s, generate SBR when invaded by an action potential. Intensive glucocorticoid dosing increases to 69% the number of frequency-conditioned terminals that generate SBR (see Table 1). This points to a decrease in the apparent threshold for SBR generation.

To probe a mechanism by which glucocorticoids may effect this change, the SBR generation requires review (see Sect. 3.2). Briefly, frequency-induced SBR is thought to arise from current flow between the preconditioned terminals and the last node of Ranvier. The preconditioning causes a subsequent hyperpolarization of unmyelinated motor nerve terminals. In this circumstance, an action potential invading the terminals has an augmented and prolonged positive afterpotential. This leads to current flow between the terminal and the node, and repetitive firing appears as a function of nodal gating properties.

From the foregoing, it is evident that if glucocorticoids were to lower the threshold of the last node, the incidence of motor axons developing SBR would be greater than normal. Since the nodal gating properties determine the characteristics of an SBR train, a decrease in nodal threshold could be predicted to: (a) decrease the latency to the appearance of the first repetitive spike, (b) decrease interspike intervals in a train, and (c) increase the number of spikes in a train. That these events occur was shown by Baker et al. (1978) (Table 4). SBR trains in motor axons of cats chronically treated with triamcinolone exhibited, on average, a 22% decrease in latency to the first spike, a 20% decrease in interspike intervals, and an increase of one spike per train (average 2.2 vs 3.2). All of these changes are significant.

A lowering of nodal threshold by intensive glucocorticoid treatment would also account for the findings of Botterman et al. (1981) with cat muscle spindle afferents. Essentially they found that defined muscle stretch activated a significantly greatly number of IA and II afferents in glucocorticoid-treated animals. These results with sensory nerve endings that are functionally dependent on a nodal excitation (cf. SBR

Table 4. Glucocorticoid effect on the pattern of stimulus-bound repetition of cat soleus motor nerve terminals conditioned by 400 Hz/10 s

	Control	Chronic triamcinolone-treated (8 mg/kg i.m./7 days)	
	ms ± SEM	ms ± SEM	
1) Mean latency to first repetitive spike	5.78 ± 0.51	3.98 ± 0.40	$p < 0.01$
2) Mean duration of individual SBR train	5.70 ± 0.20	6.20 ± 0.10	$p < 0.02$
3) Mean interspike interval in individual SBR train	3.44 ± 0.10	2.75 ± 0.37	$p < 0.01$

generation) strongly reenforce the idea that nodal excitability is importantly altered by high-dose glucocorticoid treatment.

The enhanced SBR caused by glucocorticoids focuses attention on ionic conductances. Recall that action potential generation depends on ionic currents mediated by voltage and time-dependent Na^+ and K^+ channels. Since SBR is suppressed by tetrodotoxin (TTX), in doses that do not affect axonal conduction (Baker and Riker, unpublished observations), Na^+ conductance appears to be principally altered by intensive glucocorticoid dosing. However, rigorous validation of this suggestion will require measurement of Na^+ current under voltage clamp.

Since inward current flow through Na^+ channels depends on the transmembrane voltage, unitary channel conductance, channel density, and the fraction of opened channels, the lowering of nodal threshold suggests that the fraction of Na^+ channels opened by depolarization is increased by glucocorticoids. This reasonably assumes that action potential generation does not require the opening of all Na^+ channels. With respect to this, the rigorous definition of threshold is: the voltage at which the absolute value of the inward Na^+ current exceeds that of the outward K^+ current. It is proposed, therefore, that intensive glucocorticoid dosing increases the excitability of motor axons by increasing Na^+ conductance at the nodes.

A lowering of thresholds at nodes and at spike-generating foci in other neurons would contribute to the epileptogenic action of the glucocorticoids. This action would enhance and support the spread of repetitive activity in central neurons. In this same context, the initial depressant action of the glucocorticoids should also be considered. An initial but transient action to decrease excitability (see later in this section) would tend to diminish neuronal repetitive activity. This may be the basis of glucocorticoid use in the management of myoclonic absence seizures (Eadie and Tyrer 1980). Repeated dosing, however, in view of the effects and probable mechanisms cited here, is likely to promote rather than to suppress seizures. If glucocorticoids are used to control motor seizures, a short-acting steroid such as cortisol might be the choice.

The proposal that the excitability of the spike-generating locus in susceptible neurons is increased by glucocorticoids is strengthened by the experiments of Hall (1982a, b) with cat spinal motor neurons. In addition, these experiments indicate that the sequential depressant and excitatory enhancing actions of the glucocorticoids may involve different neuronal sites. A glucocorticoid-induced depression of the soma-dendritic membrane is revealed by a rise in its firing threshold, a decrease in the rate of rise of its action potential, a decrease in overshoot, and an increase in its refractory period. On the other hand, the excitability of the axon's initial segment is enhanced by glucocorticoids. Thus, there is a 43% decrease in the current needed to evoke an action potential and a 38% decrease in the current necessary to evoke repetitive firing. In fact, the slope of the relationship between current strength and the frequency of repetitive discharge is strikingly shifted from 2.4 impulses/s/nA to 7.8 impulses/s/nA. Hall's results also reveal that glucocorticoids amplify the afterdepolarization in motor neurons and suggest that this is due to an improved invasion of the dendritic tree. Therefore, the excitability of the dendrites may be heightened.

Together these excitatory effects on motor neuron explain the effect of glucocorticoids to shift the MSR input—output relationship (Hall et al. 1978), wherein a greater number of motor neurons are caused to discharge with stimulation of a preset number of afferents. Here again the overall effect, as in motor nerve, is an increase in the number of responding units. The excitability-enhancing effect of glucocorticoids on motor neurons also reenforces the more indirect evidence, suggesting that the Ranvier nodes are a site of this action.

Since the afterdepolarization of motor neurons might involve a Ca^{2+} conductance (Barrett and Barrett 1976; Krnjevic et al. 1978), its exaggeration by glucocorticoids led Hall to suggest that Ca^{2+} channels may also be sites of glucocorticoid action. This notion is supported by the glucocorticoid-induced increase in spontaneous discharge of snail neurons (Schade and van Wilgenburg 1973), in which spike generation results in part from a voltage-dependent Ca^{2+} channel (Hagiwara and Byerly 1981). If Ca^{2+} currents in mammalian motor nerve terminals are also augmented by glucocorticoids, the observed increases in mepp frequency and in nerve-evoked endplate currents (Wilson et al. 1974; Dengler et al. 1979; van Wilgenburg 1980) would find explanation.

If direct actions of glucocorticoids on motor neurons contribute to the amelioration of myasthenia gravis, those mechanisms that appear to relate to an increased transmitter release would be the principal direct actions. This would include enhanced Ca^{2+} entry into motor nerve terminals and improved transmitter mobilization. The latter is suggested by experimental and clinical data indicating that facilitatory drug actions are enhanced by intensive glucocorticoid pretreatment. Since the traditional anticholinesterase drugs also act presynaptically to increment transmitter release and improve transmitter mobilization when these are depressed (Blaber 1973; Riker 1975), enhancement of this action may be an important aspect of glucocorticoid therapy in myasthenia.

The initial prompt suppression of SBR by glucocorticoids points to motor nerve terminal membranes as another site of action. The large surface to volume ratio of the nerve endings favors concentration of circulating substances that are soluble or reactive at this site. With respect to this, it should be noted that glucocorticoids traverse lipid membranes (Shaw et al. 1976; Fildes and Oliver 1978) and can react with the surface phospholipids (Cleary and Zatz 1977).

To suppress SBR by an action on motor nerve terminal, glucocorticoids would have to prevent or reduce the enlarged positive afterpotentials that generate the SBR. Essentially, this requires a mechanism that would reduce Na^{+} current. For example, tetrodotoxin (TTX) suppresses both high-frequency- and facilitatory-drug-induced SBR, in doses that do not affect axonal conduction (Baker and Riker, unpublished observations). However, glucocorticoids, unlike TTX, do not block axonal conduction with increasing doses. In this respect glucocorticoids would resemble phenytoin, which over a wide dose range suppresses SBR, in conjunction with a reduction in Na^{+} current, without affecting axonal conduction (Raines and Standaert 1966; Schwarz and Vogel 1977).

The seemingly persistent glucocorticoid suppression of SBR in fast motor nerves probably reflects their weak SBR generating capacity (Standaert 1963, 1964; Baker and Riker 1979). Whether this nerve terminal property recovers with time in phasic motor nerves is unknown. However, the SBR response to edrophonium appears partially recovered in 24 h (Hall 1980b). The important issue here, however, is the sharp difference that exists between the glucocorticoid responsiveness of two neurons closely related functionally. Considering that these contrasting direct sensitivities could occur in CNS neurons, in conjunction with other regional differences, it is not surprising that responses to locally applied glucocorticoids may be difficult to interpret.

The glucocorticoid depressant action can also be differentiated from its excitatory action by experiments (Baker and Riker, unpublished observations) which show that 11β-cortisol but not 11α-cortisol augments SBR generation. In contrast, both epimers are equipotent in causing the prompt and transient initial depression of SBR. The significance of the structural specificity for cortisol (and presumably other glucocorticoids) enhancement of excitability is unclear.

How the unmyelinated motor nerve terminal membranes and other susceptible neuronal regions might be altered by glucocorticoids can only be surmised. Conceivably, the depressant action of these steroids could arise from membrane disturbances like those caused by a variety of nonspecifically acting depressants (e.g., general anesthetics; for review see Seeman 1972; Roth 1979). Drugs of this type can disorder the membrane's hydrophobic lipid core, increase lipid fluidity, and effect conformational changes in membrane proteins. These changes also cause both model and biologic membranes to expand in volume up to some ten times that which is attributable to solvation. It has been suggested, therefore, that these membrane changes lead to a loss of excitability, perhaps by impairing ion channel openings.

Although glucocorticoids traverse lipid membranes, and in so doing may incur the aforementioned membrane disorders, their retention in the hydrophobic lipid bilayer is improbable. Cortisol can be incorporated into phospholipid bilayers but it is not retained (Shaw et al. 1976; Fildes and Oliver 1978). Liposomes release cortisol and its 21-succinate, 21-acetate, and 21-phosphate esters within 24 h (Shaw et al. 1976). Therefore, the initial depressant action of glucocorticoids might result from a transient disordering associated with their initial membrane traversal. That they are not retained by lipid may account for their relatively short-lived depressant action.

The disappearance of their depressant action with continued treatment may reflect a change in membrane composition at these sites of glucocorticoid action. There is good evidence from experiments with both model and neuronal membranes to suggest that this alteration may be determined by changes in membrane cholesterol content. Thus, Snart and Wilson (1967), Cleary and Zatz (1977), and Fildes and Oliver (1978) found that cholesterol incorporation into phospholipid membranes or liposomes either prevented glucocorticoid penetration to the surface of the membrane or speeded its expulsion. These findings suggest that regions of neuronal membranes having a low cholesterol content may be more susceptible to glucocorticoid action, and vice versa. Perhaps the unmyelinated motor nerve terminals, the axon hillock, Ranvier nodes, and primary afferent endings are examples.

That this might be the case has been disclosed by recent experiments showing that membrane cholesterol content alters excitability. Nakajima and Bridgman (1981), using electron microscopic freeze fracture and a cholesterol-specific reagent (filipin), found for frog neuromuscular junction a low cholesterol content at the active zones of presynaptic membranes and at the receptor-rich tips of the postjunctional folds. This suggests that it is functionally advantageous to keep cholesterol content low at membrane areas having a high ion channel activity.

The reports of Stephens and Shinitzky (1977) and of Redmann et al. (1979) provide important information. They describe the effects of cholesterol on the excitability of neurons in vitro. Their results encourage consideration of membrane cholesterol content as a determinant of susceptibility to glucocorticoids. Thus, cholesterol enrichment of *Aplysia* ganglion cells in vitro, via cholesterol-containing liposomes, inhibited their spontaneous discharge, eventually stopping it (Stephens and Shinitzky 1977). Reduction of ganglion cell cholesterol, via exposure to non-cholesterol-containing liposomes, both restored and speeded their spontaneous discharge. In voltage-clamped squid axons (Redmann et al. 1979), cholesterol enrichment, mediated by cholesterol-lecithin liposomes, reduced and slowed the peak inward Na^+ current. The findings of these workers show that respectively increasing and decreasing cholesterol membrane content mimics the biphasic effects of the glucocorticoids on excitability. Therefore, the transience of the initial glucocorticoid SBR suppression could result from the expulsion of the glucocorticoid from a phospholipid membrane, after which an adaptive increase in the cholesterol content would prevent further glucocorticoid penetration.

It is conceivable that a glucocorticoid condensing effect (Cleary and Zatz 1977) might alter ion channel "accessibility." For example, the idea of movement of membrane proteins was studied with respect to cholesterol by Borochov and Shinitzky (1976). Increasing the membrane fluidity of erythrocytes via cholesterol reduction "masked" the membrane proteins. Conversely, decreasing fluidity, i.e., increasing order, by cholesterol addition increased the exposure of these membrane proteins to the aqueous surround. Although glucocorticoids do not interact with lipid membranes like cholesterol (for cholesterol interaction see Huang 1977), they may, via their interaction with phospholipid cationic polar head groups, increase membrane ordering and in turn alter the properties of ion channel proteins.

In sum, there is now abundant evidence showing that glucocorticoids in high doses can directly alter the excitability of neuronal membranes bidirectionally. These effects are complex in that they clearly involve multiple neuronal sites and follow different time courses. They are further complicated by differing neuronal susceptibilities, even in neurons that are closely related functionally (e.g., motoneurons innervating fast and slow muscles). Most of these glucocorticoid effects exhibit characteristics representative of nonspecific drug action on membranes. Insight into ways in which glucocorticoids might interact with neuronal membranes to alter function is frustrated by the paucity of studies on glucocorticoid interactions with model and biologic membranes. On the other hand, an important role of cholesterol in excitable membranes is emerging. In the light of what is known of glucocorticoid–cholesterol interactions in lipid membranes, the cholesterol content of neuronal membranes might prove to be a critical factor in determining neuronal susceptibility to glucocorticoid actions, as well as the nature of the neuronal response. Finally, the direct membrane actions of glucocorticoids should not be overlooked in considering the clinical consequences of intensive glucocorticoid treatment. The neurologic and behavioral changes caused by high-dose glucocorticoids may arise in part from such actions; so too may certain of the beneficial effects of glucocorticoids in the treatment of myasthenia gravis.

Acknowledgments. This review, as well as certain of the cited research from the authors' laboratories, was supported by United States Public Health Service, National Institute of Neurological Communicative Disorders and Stroke Grant 01447. The authors express their gratitude to Richard Zwies and Mary Lou Steger for their invaluable assistance.

References

Arts WF, Oosterhuis HJ (1975) Effect of prednisolone on neuromuscular blocking in mice in vivo. Neurol 25:1088–1090

Arts WF, Oosterhuis HJ (1977) Long-term effect of glucocorticoids on neuromuscular blocking in mice. J Neurol Neurosurg Psychiat 40:675–677

Astwood EB, Cleroux AP, Payne RW, Raben MS (1950a) Therapeutic studies on some newer corticotrophic (ACTH) preparations. Bull N Engl Med Ctr 12:2–10

Astwood EB, Raben MS, Payne RW, Cleroux AP (1950b) Clinical evaluation of crude and highly purified preparations of corticotrophin (ACTH) obtained in good yields by simple laboratory procedures. J Clin Invest 29:797

Baker T, Riker WF Jr (1979) Pharmacologic differences between cat tonic and phasic motor nerve terminals. Pharmacologist 21:244

Baker T, Riker WF Jr, Hall ED (1977) Effects of a single methylprednisolone dose on a facilitatory response of mammalian motor nerve. Arch Neurol 34:349–355

Baker T, Riker WF Jr, Hall ED (1978) Triamcinolone enhancement of cat soleus motor nerve terminal excitability. Fed Proc 37:577

Barak Y Ben, Gutnick MJ, Feldman S (1977) Iontophoretically applied corticosteroids do not affect the firing of hippocampal neurons. Neuroendocrinol 23:248–256

Barrett EF, Barrett JN (1976) Separation of two voltage-sensitive potassium currents, and demonstration of a tetrodotoxin-resistant calcium current in frog motoneurones. J Physiol (Lond) 255:737–774

Blaber LC (1973) The prejunctional actions of some non-depolarizing blocking drugs. Br J Pharmacol 47:109–116

Borochov H, Shinitzky M (1976) Vertical displacement of membrane proteins mediated by changes in microviscosity. Proc Natl Acad Sci USA 73:4526–4530

Botterman BR, Eldred E, Edgerton VR (1981) Spindle discharge in glucocorticoid-induced muscle atrophy. Exp Neurol 72:25–40

Bowman WC, Rand MJ (1980) Textbook of pharmacology, 2nd edn. Blackwell, London

Bowman WC, Webb SN (1976) Tetanic fade during partial transmission failure produced by non-depolarizing neuromuscular blocking drugs in the cat. Clin Exp Pharmacol Physiol 3:545–555

Brunner NG, Namba T, Grob D (1972) Corticosteroids in management of severe, generalized myasthenia gravis: Effectiveness and comparison with corticotropin therapy. Neurology 22:603–610

Brunner NG, Berger CL, Namba T, Grob D (1976) Corticotropin and corticosteroids in generalized myasthenia gravis: Comparative studies and role in management. Ann NY Acad Sci 274:577–595

Cleary GW, Zatz JL (1977) Interaction of hydrocortisone with model membranes containing phospholipid and cholesterol. J Pharmaceutical Sci 66:975–980

Covian MR, Lico MC, Antunes-Rodrigues J (1963) Effects of adrenal corticoids on visual cortical potentials in the cat. Arch Int Pharmacodyn Ther 146:81–92

Dameshek W, Saunders RH, Zannos L (1950) Use of ACTH in treatment of acute and subacute leukemia. Bull N Engl Med Ctr 12:11–21

Davidson JM, Feldman S (1963) Cerebral involvement in the inhibition of ACTH secretion by hydrocortisone. Endocrinology 72:936–946

Dengler R, Rudel R, Warelas J, Birnberger KL (1979) Corticosteroids and neuromuscular transmission: electrophysiological investigation of the effects of prednisolone on normal and anticholinesterase-treated neuromuscular junction. Pfluegers Arch Eur J Physiol 380:145–151

Dlouha H, Vyskocil F (1979) The effect of cortisol on the excitability of the rat muscle fibre membrane and neuromuscular transmission. Physiologia Bohemoslov 28:485–494

Dorfman A, Apter NS, Smull K, Bergenstal DM, Richter RB (1951) Status epilepticus coincident with use of pituitary adrenocorticotropic hormone: Report of three cases. JAMA 146:25–27

Drachman DB (1978) Myasthenia gravis (Part II). N Engl J Med 298:186–193

Dreyer F, Peper K, Sterz R, Bradley RJ, Muller KD (1979) Drug receptor interaction at the frog neuromuscular junction. Progr Brain Res 49:213–223

Dudel J, Birnberger KL, Toyka KV, Schlegel C, Besinger U (1979) Effects of myasthenic immunoglobulins and of prednisolone on spontaneous miniature end-plate potentials in mouse diaphragms. Exp Neurol 66:365–380

Eadie MJ, Tyrer JM (1980) Anticonvulsive therapy, pharmacological basis and practice, 2nd edn. Churchill Livingstone, Edinburgh New York

Endroczi E, Koranyi L (1968) Integration of emotional reactions in brain-stem, diencephalic and limbic system. In: Gara Hini S, Sigg EB (eds) Aggressive behavior: Proceedings of the International Symposium on the Biology of Aggressive Behavior. Wiley & Sons, New York, pp 132–140

Engel WK (1976) Myasthenia gravis, corticosteroids, anticholinesterases. Ann NY Acad Sci 274:623–630

Feldman S, Dafny N (1970a) Effects of cortisol on unit activity in the hypothalamus of the rat. Exp Neurol 27:375–387

Feldman S, Dafny N (1970b) Changes in single cell responsiveness in the hypothalamus in cats following cortisol administration. Brain Res 20:369–377

Feldman S, Todt JC, Porter RW (1961) Effect of adrenocortical hormones on evoked potentials in the brain stem. Neurology 11:109–115

Feng TP (1938) Studies on the neuromuscular junction. X. The effects of guanidine. Chin J Physiol 13:119–140

Fildes FJT, Oliver JE (1978) Interaction of cortisol-21-palmitate with liposomes examined by differential scanning calorimetry. J Pharm Pharmacol 30:337–342

Freydberg LD (1960) The place of corticotrophin in the treatment of myasthenia gravis. Ann Intern Med 52:806–818

Galindo A (1971) Prejunctional effect of curare: Its relative importance. J Neurophysiol 34:289–301

Gammon GD, Scheie H (1937) Use of prostigmin as a diagnostic test of myasthenia gravis. JAMA 109:413–414

Gandiglio G, Pinelli P (1968) EMG analysis of myasthenic ACTH deterioration. Eur Neurol 1:2–16

Glaser GH (1953) On the relationship between adrenal cortical activity and the convulsive state. Epilepsia (Third Series) 2:7–14

Glaser GH, Kornfeld DS, Knight RP (1955) Intravenous hydrocortisone, corticotrophin and electro-encephalogram. Arch Neurol Psychiatr 73:338–344

Glavinovic MI (1979) Change of statistical parameters of transmitter release during various kinetic tests in unparalysed voltage-clamped rat diaphragm. J Physiol (Lond) 290:481–497

Goodman LS, Gilman A (1980) The pharmacological basis of therapeutics, 6th edn. MacMillan, New York

Grob D, Harvey AM (1952) Effect of adrenocorticotropic hormone (ACTH) and cortisone administration in patients with myasthenia gravis and report of onset of myasthenia gravis during prolonged cortisone administration. Bull Johns Hopkins Hosp 91:124–136

Grob D, Namba T (1966) Corticotropin in generalized myasthenia gravis. JAMA 198:703–707

Grossie J, Albuquerque EX (1978) Extensor muscle response to triamcinolone. Exp Neurol 58:435–445

Hagiwara S, Byerly L (1981) Calcium channel. Ann Rev Neurosci 4:69–125

Hall ED (1980a) Glucocorticoid enhancement of guanidine neuromuscular facilitation. Exp Neurol 68:581–588

Hall ED (1980b) Glucocorticoid modification of the responsiveness of a fast (Type 2) neuromuscular system to edrophonium and d-tubocurarine. Exp Neurol 69:349–358

Hall ED (1982a) Glucocorticoid effect on the electrical properties of spinal motor neurons. Brain Res

Hall ED (1982b) Acute effects of intravenous glucocorticoid on cat spinal motor neuron electrical properties. Brain Res

Hall ED, Baker T (1979a) Acute effects of methyl prednisolone sodium succinate on spinal reflexes. Exp Neurol 63:476–484

Hall ED, Baker T (1979b) Further studies of glucocorticoid effects on spinal cord function: single and repetitive monosynaptic transmission and apparent Ia afferent transmitter turnover. J Pharmacol Exp Ther 210:112–115

Hall ED, Baker T, Riker WF Jr (1977a) Glucocorticoid effects on the edrophonium responsiveness of normal and degenerating mammalian motor nerve terminals. Ann Neurol 2:404–408

Hall ED, Baker T, Riker WF Jr (1977b) Glucocorticoid effects on motor nerve function during early degeneration. Ann Neurol 1:263–269

Hall ED, Baker T, Riker WF Jr (1978) Glucocorticoid effects on spinal cord function. J Pharmacol Exp Ther 206:361–370

Harvey AM, Lilienthal JL Jr (1941) Observations on the nature of myasthenia gravis. The intraarterial injection of acetylcholine, prostigmine and adrenaline. Bull Johns Hopkins Hosp 69:566–577

Harvey AM, Lilienthal JL Jr, Talbot SA (1941) On the effects of the intraarterial injection of ACh and prostigmine in normal man. Bull Johns Hopkins Hosp 69:529–546

Hicks SP (1953) Developmental brain metabolism. Effects of cortisone, anoxia, fluoroacetate, radiation, insulin and other inhibitors on the embryo, newborn and adult. Arch Pathol 55:302–327

Hoagland H, Bergen JR, Slocombe AG, Hunt CA (1953) Studies of adrenocortical physiology in relation to the nervous system. Res Publ Assoc Res Nerv Dis 32:40–60

Hoffmann WW (1977) Antimyasthenic action of corticosteroids. Arch Neurol 34:356–360

Holtkamp DE, Mansor LF, Herning HE (1952) Electroshock threshold and dose response to cortisone administration. Fed Proc 11:358

Howard FM Jr, Duane DD, Lambert EH, Daube JR (1976) Alternate-day prednisolone: preliminary report of a double-blind controlled study. Ann NY Acad Sci 274:596–607

Huang C-H (1977) A structural model for the cholesterol-phosphatidylcholine complexes in bilayer membranes. Lipids 12:348–356

Hubbard JI, Schmidt RF (1963) An electrophysiological investigation of mammalian motor nerve terminals. J Physiol (Lond) 166:145–167

Hubbard JI, Wilson DF (1973) Neuromuscular transmission in a mammalian preparation in the absence of blocking drugs and the effect of d-tubocurarine. J Physiol (Lond) 228:307–325

Hubbard JI, Schmidt RF, Yokota T (1965) The effect of acetylcholine upon mammalian motor nerve terminals. J Physiol (Lond) 181:810–829

Hubbard JI, Wilson DF, Miyamoto M (1969) Reduction of transmitter release by d-tubocurarine. Nature 223:531–532

Jenkins RB (1972) Treatment of myasthenia gravis with prednisone. Lancet I:765–767

Johns TR (1977) Treatment of myasthenia gravis: Long-term administration of corticosteroids with remarks on thymectomy. Adv Neurol 17:99–102

Kamenskaya MA, Elmqvist D, Thesleff S (1975a) Guanidine and neuromuscular transmission I. Effect on transmitter release occurring spontaneously and in response to single nerve stimuli. Arch Neurol 32:505–509

Kamenskaya MA, Elmqvist D, Thesleff S (1975b) Guanidine and neuromuscular transmission. II. Effect on transmitter release in reponse to repetitive nerve stimulation. Arch Neurol 32:510–518

Katz B, Miledi R (1965) Release of acetylcholine from a nerve terminal by electric pulses of variable strength and duration. Nature 207:1097–1098

Kelly MJ, Moss RL, Dudley CA (1977) The effects of microelectrophoretically applied estrogen cortisol and acetylcholine on medial preoptic septal unit activity throughout the estrous cycle of the female rat. Exp Brain Res 30:53–64

Kim YI, Goldner MM, Sanders DB (1979) Short-term effects of prednisolone on neuromuscular transmission in normal rats and those with experimental autoimmune myasthenia gravis. J Neurol Sci 41:223–234

Kjaer M (1971) Myasthenia gravis and myasthenic syndromes treated with prednisone. Acta Neurol Scand 47:464–474

Koranyi L, Endroczi E (1970) Influence of pituitary-adrenocortical hormones on thalamo-cortical and brain stem limbic circuits. Prog Brain Res 32:120–130

Krnjevic K, Lamour Y, MacDonald JF, Nistri A (1978) Motoneuronal afterpotentials and extracellular divalent cations. Can J Phys Pharm 56:516–520

Leeuwin RS, Wolters EChMJ (1977) Effects of corticosteroid on sciatic nerve-tibialis anterior muscle of rats treated with hemicholinium-3. Neurol (Minn) 27:171–177

Leeuwin RS, Veldsema-Currie RD, Wolters EChMJ (1978) The effects of cholinesterase inhibitors and corticosteroids on rat nerve-muscle preparations treated with hemicholinium-3. Eur J Pharmacol 50:393–401

Liley AW, North KAK (1953) An electrical investigation of effects of repetitive stimulation on mammalian neuromuscular junction. J Neurophysiol 16:509–527

Lloyd DPC (1949) Post-tetanic potentiation of response in monosynaptic reflex pathways of the spinal cord. J Gen Physiol 33:147–170

Lloyd DPC, McIntyre AK (1949) On the origins of dorsal root potentials. J Gen Physiol 32:409–443

Maeno T (1969) Analysis of mobilization and demobilization processes in neuromuscular transmission in the frog. J Neurophysiol 32:793–800

Maeno T, Edwards C (1969) Neuromuscular facilitation with low-frequency stimulation and effects of some drugs. J Neurophysiol 32:785–792

Magleby KL, Pallotta BS, Terrar DA (1981) The effect of (+)-tubocurarine on neuromuscular transmission during repetitive stimulation in the rat, mouse and frog. J Physiol 312:97–113

Mangili G, Motta M, Martini L (1966) Control of adrenocorticotrophic hormone secretion. In: Martini L, Ganong WE (eds) Neuroendocrinology. Academic Press, New York, pp 297–370

Mann JD, Johns TR, Campa JF, Muller WH (1976) Long-term prednisone followed by thymectomy in myasthenia gravis. Ann NY Acad Sci 274:608–622

Mansor LF, Holtkamp DE, Heming AE, Christain HH (1956) Effects of prednisone, prednisolone and fludrocortisone acetate on electroshock seizure threshold. Fed Proc 15:454

Masland RL, Wigton RS (1940) Nerve activity accompanying fasciculations produced by prostigmin. J Neurophysiol 3:269–275

McEwen BS (1981) Neuronal gonadal steroid actions. Science 211:1303–1311

McEwen BS, Davis PG, Parsons B, Pfaff DW (1979) The brain as a target for steroid hormone action. Ann Rev Neurosci 2:65–112

Millikan CH, Eaton LM (1951) Clinical evaluation of ACTH and cortisone in myasthenia gravis. Neurol (Minn) 1:145–152

Minot AN, Dodd K, Riven SS (1939) Use of guanidine hydrochloride on transmitter release and neuronal excitability. JAMA 113:553–559

Misrahy G, Toman JEP (1953) Effects of drugs on irradiation of cerebral cortical electrical responses in the rabbit. Fed Proc 12:352

Miyamoto MD (1977) The actions of cholinergic drugs on motor nerve terminals. Pharmacol Rev 29:221–247

Nakajima Y, Bridgman PC (1981) Absence of filipin-sterol complexes from the membranes of active zones and acetylcholine receptor aggregates at frog neuromuscular junction. J Cell Biol 88:453–458

Namba T (1972) Corticotropin therapy in patients with myasthenia gravis: Electrophysiologic, pharmacologic studies. Arch Neurol 26:144–150

Namba T, Shapiro MS, Arimori S, Grob D (1967) Effects of corticotropin in patients with generalized myasthenia gravis. J Clin Invest 46:1100

Namba T, Brunner NG, Shapiro MS, Grob D (1971) Corticotrophin therapy in myasthenia gravis: Effects, indications and limitations. Neurol (Minn) 21:1008–1018

Oh SJ, Kim KW (1973) Guanidine hydrochloride in the Eaton-Lambert syndrome: electrophysiological improvement. Neurol 23:1084–1090

Osserman KE, Genkins G (1966) Studies in myasthenia gravis. JAMA 198:699–702

Osserman KE, Kaplan LI (1952) Rapid diagnostic test for myasthenia gravis. JAMA 150:265–268

P'an SY, Laubach GD (1964) Steroid central depressants. In: Dorfman RJ (ed) Methods in hormone research, vol 3. Steroidal activity in experimental animals and man. Part A. Academic Press, New York, pp 415–475

Pasolini F (1952) Epilepsy and endocrine glands: II. Influence of cortisone on predisposition to experimental reflex epilepsy in the dog. Boll Soc Ital Biol Sper 28:298–300

Patten BM, Oliver KL, Engel WK (1974) Adverse interaction between steroid hormones and anticholinesterase drugs. Neurol (Minn) 24:442–449

Pfaff DW, Silvia MTA, Weiss JM (1971) Telemetered recordings of hormone effects on hippocampus neurons. Science 172:394–395

Pincus JB, Natelson S, Lugovoy JK (1951) Effect of epinephrine, ACTH and cortisone on citrate, calcium, glucose and phosphate levels in rabbits. Proc Soc Exp Biol Med 78:24–27

Pinelli P, Tonali P, Scoppetta C (1974) Long-term treatment of myasthenia gravis with alternate-day prednisone. Eur Neurol 12:129–141

Raines A, Standaert FG (1966) Pre- and post-junctional effects of diphenylhydantoin at the cat soleus neuromuscular junction. J Pharmacol Exp Ther 153:361–366

Redmann GA, Brodwick MS, Eaton DC, Steele J, Poznansky M (1979) Liposome mediated lipid exchange effects on squid axon membrane currents (Abstr). Biophys J 25:138a

Riker DK, Sastre A, Baker T, Roth RH, Riker WF Jr (1979) Regional high-affinity [^{3}H] choline accumulation in cat forebrain: selective increase in the caudate-putamen after corticosteroid pretreatment. Mol Pharmacol 16:886–899

Riker WF Jr (1966) The motor nerve terminal as a pharmacologic substrate. In: Costa E, Cote LJ, Yahr MD (eds) Biochemistry and pharmacology of the basal ganglia. Proceedings of the Second Symposium of the Parkinson's Disease Information and Research Center. Raven Press, New York, pp 43–63

Riker WF Jr (1975) Prejunctional effects of neuromuscular blocking and facilitatory drugs. In: Katz RL (ed) Muscle relaxants. North Holland, Amsterdam, pp 59–102

Riker WF, Standaert FG (1966) The actions of facilitatory drugs and acetylcholine on neuromuscular transmission. Ann NY Acad Sci 135:163–176

Riker WF Jr, Okamoto M (1969) Pharmacology of motor nerve terminals. Ann Rev Pharmacol 9:173–208

Riker WF, Werner G, Roberts J, Kuperman A (1959) Pharmacologic evidence for the existence of a presynaptic event in neuromuscular transmission. J Pharmacol Exp Ther 125:150–158

Riker WF Jr, Baker T, Okamoto M (1975) Glucocorticoids and mammalian motor nerve excitability. Arch Neurol 32:688–694

Riker WF Jr, Hall ED, Baker T (1977) Glucocorticoid unmasking of an anticurare action of curare. Pharmacologist 19:197

Rosenbaum RB, Bender AN, Engel WK (1975) Prolonged response to edrophonium in myasthenia gravis. Trans Am Neurol Assoc 100:233–235

Roth SH (1979) Physical mechanisms of anesthesia. Ann Rev Pharmacol Toxicol 19: 159–178

Schade JP, van Wilgenburg H (1973) Steroid sensitivity of the nervous system. In: Lissak K (ed) Hormones and brain function. Plenum Press, New York, pp 375–378

Schlezinger NS (1952) Present status of therapy in myasthenia gravis. JAMA 148: 508–513

Schwarz JR, Vogel W (1977) Diphenylhydantoin: excitability reducing action in single myelinated nerve fibers. Eur J Pharmacol 44:241–249

Seeman P (1972) The membrane actions of anesthetics and tranquilizers. Pharmacol Rev 24:583–655

Seybold ME, Drachman DB (1974) Gradually increasing doses of prednisone in myasthenia gravis: reducing the hazards of treatment. N Engl J Med 290:81–84

Shaw IH, Knight GC, Dingle JT (1976) Liposomal retention of a modified anti-inflammatory steroid. Biochem J 158:473–476

Shy GM, Brendler S, Rabinovitch R, McEachern D (1950) Effects of cortisone in certain neuromuscular disorders. JAMA 144:1353–1358

Simon HE (1935) Myasthenia gravis: Effect of treatment with anterior pituitary extract: preliminary report. JAMA 104:2065–2066

Slusher MA, Hyde JE, Laufer M (1966) Effect of intracerebral hydrocortisone on unit activity of diencephalon and midbrain in cats. J Neurophysiol (Lond) 29:157–169

Snart RS, Wilson MJ (1967) Uptake of steroid hormones into artificial phospholipid cholesterol membranes. Nature 215:964

Standaert FG (1963) Post-tetanic repetitive activity in the cat soleus nerve. J Gen Physiol 47:53–70

Standaert FG (1964) The mechanisms of post-tetanic potentiation in cat soleus and gastrocnemius muscles. J Gen Physiol 47:987–1001

Standaert FG, Riker WF Jr (1967) The consequences of cholinergic drug actions on motor nerve terminals. Ann NY Acad Sci 144:517–533

Steiner FA (1970) Effects of ACTH and corticosteroids on single neurons in the hypothalamus. Prog Brain Res 32:102–107

Stephen EHM, Noad KB (1951) Status epilepticus occurring during cortisone therapy. Med J Australia 2:334–335

Stephens CL, Shinitzky M (1977) Modulation of electrical activity in *aplysia* neurones by cholesterol. Nature 270:167–268

Teravainen H, Larsen A (1975) Effect of guanidine on quantal release of acetylcholine in mammalian myoneuronal junction. Exp Neurol 48:601–609

Teyler TJ, Vardaris RM, Lewis D, Rawitch AB (1980) Gonadal steroids: Effects on excitability of hippocampal pyramidal cells. Science 209:1017–1019

Timiras PS, Woodbury DM, Goodman LS (1954) Effect of adrenalectomy, hydrocortisone acetate and desoxycorticosterone acetate on brain excitability and electrolyte distribution in mice. J Pharmacol Exp Ther 112:80–93

Timiras PS, Woodbury DM, Baker DH (1956) Effect of hydrocortisone acetate, desoxycorticosterone acetate, insulin, glucagon and dextrose, alone or in combination on experimental convulsions and carbohydrate metabolism. Arch Int Pharmacodyn Ther 105:450–467

Torda C, Wolff HC (1949) Effects of adrenocorticotrophic hormone on neuromuscular function in patients with myasthenia gravis. Proc Soc Exp Biol Med 71:432–435

Torda C, Wolff HG (1951) Effects of administration of adrenocorticotrophic hormone on patients with myasthenia gravis. Arch Neurol Psychiatr 66:163–170

Towle AC, Sze PY (1978) Binding of corticosterone to synaptic plasma membrane from rat brain (Abstr). Soc Neurosci 4:356

Turner BB, DeFiore C (1980) Existence of glucocorticoid binding in the striatum: mechanism for increased choline uptake. Physiologist 23:144

van Wilgenburg H (1979) The effect of prednisolone on neuromuscular transmission in the rat diaphragm. Eur J Pharmacol 55:355–361

van Wilgenburg H (1980) Effects of glucocorticoids on acetylcholine release at the neuromuscular junction. Br J Pharmacol 68:144

Veldsema-Currie RD, Wolters EChMJ, Leeuwin RS (1976) The effect of corticosteroids and hemicholinium-3 on choline uptake and its incorporation into acetylcholine in rat diaphragm. Eur J Pharmacol 35:399–402

Volle RL (1966) Enhancement of neuromuscular responses to acetylcholine and succinylcholine following repetitive stimulation of the motor nerve. Arch Int Pharmacodyn Ther 160:284–293

von Reis G, Liljestrand A, Matell G (1965) Results with ACTH and spironolactone in severe cases of myasthenia gravis. Acta Neurol Scand (Suppl) 41:463–471

Werner G (1960) Neuromuscular facilitation and antidromic discharges in motor nerves: their relation to activity in motor nerve terminals. J Neurophysiol 23:171–187

Wilson RW, Ward MD, Johns TR (1974) Corticosteroids: a direct effect at the neuromuscular junction. Neurology 24:1091–1095

Withrow CD, Woodbury DM (1964) Direct and indirect effects of desoxycorticosterone (DOC) on skeletal muscle electrolyte and acid-base metabolism. In: Martini L, Pecile A (eds) Hormonal steroids, biochemistry, pharmacology and therapeutics. Proceedings of the First International Congress on Hormonal Steroids, vol 1. Academic Press, New York, pp 503–513

Withrow CD, Woodbury DM (1972) Some aspects of the pharmacology of adrenal steroids and the central nervous system. In: Reulen HJ, Schürmann K (eds) Steroids and brain edema. Springer, Berlin Heidelberg New York, pp 41–55

Wolters EChMJ, Leeuwin RS (1975) Antagonism of corticosteroids against the lethal effects of hemicholinium-3 in rats and mice. Eur J Pharmacol 33:145–149

Wolters EC, Leeuwin RS (1976) Effects of corticosteroids on the phrenic nerve-diaphragm preparation treated with hemicholinium. Neurol (Minn) 26:574–578

Woodbury DM (1952) Effect of adrenocortical steroids and adrenocorticotrophic hormones on electroshock threshold. J Pharmacol Exp Ther 105:27–36

Woodbury DM (1954) Effect of hormones on brain excitability and electrolytes. Recent Prog Horm Res 10:65–107

Woodbury DM (1958) Relation between the adrenal cortex and the central nervous system. Pharmacol Rev 10:275–357

Woodbury DM (1972) Biochemical effects of adrenocortical steroids on the nervous system. In: Laitha A (ed) Handbook of neurochemistry, vol VII, Pathological chemistry of the nervous system. Plenum Press, New York, pp 255–287

Woodbury DM, Vernadakis A (1966) Effects of steroids on the central nervous system. In: Methods in hormone research, vol 5. Steroidal activity in experimental animals and man, part C. Academic Press, New York, pp 1–57

Zucker RS (1973) Changes in the statistics of transmitter release during facilitation. J Physiol (Lond) 229:787–810

Adrenal Steroids and Behavioral Adaptation: Relationship to Brain Corticoid Receptors

BÉLA BOHUS, E.R. DE KLOET, and H.D. VELDHUIS[1]

Contents

---

1 Rudolf Magnus Institute for Pharmacology, Medical Faculty, University of Utrecht, Vondellaan 6, 3521 GD Utrecht, The Netherlands

1 Introduction

The words adrenocortical hormones and adaptation have been linked for almost 40 years. The concept as formulated by Selye (1950) related the adrenal steroids to physiological, primarily peripheral adaptive processes. It has, however, become evident that the brain is an important target organ for corticosteroids. Psychological changes, such as mood alterations and psychotic reactions, have been frequently observed in patients receiving corticosteroid therapy (see von Zerssen 1976). Although the majority of reports hardly exceeded anecdotal level, therapeutic hopes were stirred for the mentally ill. The hopes and wishes of the early 1950s remained unfulfilled. Regular basic research concerning the pituitary-adrenal system and behavior was not started until the 1960s. Concepts which stemmed from this research led to the endocrine view of brain function and dysfunction held today. In this paper we want to review the past (not seeking completeness), to describe the present state of research, and to provide our view of the future of corticosteroid-brain interactions in understanding the endocrine modulation of behavioral adaptation.

2 Behavioral Effects of Adrenalectomy and Corticosteroids

2.1 Adrenalectomy

Removal of the adrenals does not impair the acquisition of fear-motivated (avoidance) behavior of the rat provided that the salt and water balance is maintained with sodium chloride in drinking water or by mineralocorticoid administration (Fuller et al. 1956; Moyer 1958; Moyer and Moshein 1963; Bohus and Endröczi 1965; de Wied 1967; Bohus and Lissak 1968; van Delft 1970; Weiss et al. 1970). Under certain conditions, such as high shock intensity punishment, adrenalectomy improves avoidance acquisition by attenuating the deleterious effect of strong punishment (Beatty et al. 1970). Appetitive maze learning, especially during the early phase, is lightly enhanced in adrenalectomized rats (Paul and Havlena 1962).

Adrenalectomy interferes profoundly with the maintenance of avoidance behaviors in the absence of punishment. Active avoidance extinction or passive avoidance retention tests are the conventionally used measures of response maintenance. Adrenalectomized rats are resistant to extinction (de Wied 1967; Bohus et al. 1968; Weiss et al. 1970; Silva 1974) and display facilitated passive avoidance retention (Weiss et al. 1970; Silva 1974).

The following abbreviations appear in the text:

ACTH: adrenocorticotropic hormone; ADX: adrenalectomy; Aldo: aldosterone; B: corticosterone; B max: maximal binding capacity; Dex: dexamethasone; Doc: deoxycorticosterone; DRL: differential reinforcement of low rate response; F: cortisol; hypox: hypophysectomy; Kd: dissociation constant; Prog: progesterone; RBA: relative binding affinities; SAX: sham adrenalectomy; TL: transcortin-like.

These observations seem to suggest that absence of circulating adrenal steroids improves rather than impairs acquisition and maintenance of learned behaviors. It should, however, be emphasized that these studies have been performed after a relatively long time (1 week or more) following adrenal removal. Pituitary ACTH synthesis and release are substantially increased after adrenalectomy, but following a few days of lag period (Buckingham and Hodges 1974; Dallman et al. 1974). Increased ACTH or related peptide levels by peripheral administration result in a slight improvement of acquisition behavior and resistance to extinction both in adrenalectomized and intact rats (for review see Bohus 1979). It is therefore more likely that an increased ACTH rather than a decreased adrenal steroid level is responsible for the afore-mentioned behavioral alterations.

Our recent observations favor this idea. Rats trained under mild punishment conditions for a one-trial learning passive avoidance response display retention deficit if the adrenals have been removed 1 or 2 days before learning. This behavioral deficit disappears if the training takes place 10 days after adrenal removal. Corticosterone supplementary therapy does not normalize passive avoidance behavior of the adrenalectomized rats. Adrenomedullectomy causes similar deficit, which can be restored by small amounts of adrenaline or noradrenaline. Accordingy, learning and/or retention deficit is caused by the impaired adrenomedullary rather than adrenocortical function. Normal behavior of adrenalectomized, but not of medullectomized rats 10 days after surgery suggests that ACTH hypersecretion may compensate for the absence of medullary catecholamines in terms of behavioral effects (Borrell et al., in preparation).

Prevention of experimentally induced memory loss (amnesia) by long-term adrenalectomy also agrees with the suggestion that increased production of pituitary ACTH rather than absence of steroids is the primary factor in the majority of behavioral changes. Electrically or chemically induced brain seizures, protein synthesis inhibitors, or kindled convulsion from the amygdala, administered shortly after learning, result in retrograde amnesia for the experience (McGaugh et al. 1972; Dunn 1980). Amnesia induced by puromycin (Flexner and Flexner 1970), cycloheximide (Nakajima 1975), kindled amygdaloid convulsion (McIntyre 1976; McIntyre and Wann 1978), and electroconvulsive shock (Bookin and Pfeifer 1978) are prevented by long-term adrenalectomy. Administration of ACTH fragments leads to similar antiamnesic effects in intact rats (Rigter and Crabbe 1979).

2.2 Effects of Corticosteroids on Learning, Performance, and Retention in Intact Rats

Observations on the behavioral effects of corticosteroids in intact rats have been performed mainly with the easily accessible, potent glucocorticoids, such as cortisone, cortisol, or dexamethasone. Corticosterone, which is the main secretory product of the rat adrenal cortex, has been used primarily from the early seventies. In addition, the applied doses have been in the milligram range except for the highly potent synthetic glucocorticoid dexamethasone.

Learning of active avoidance responses (multiple punishment) is not affected by glucocorticoids like cortisone (Bohus and Lissak 1968) or dexamethasone (Conner and Levine 1969; Beatty et al. 1970). However, passive avoidance learning (single or few punishment) and conditioned taste aversion learning are attenuated by various glucocorticoids (Bohus et al. 1970; Bohus 1971, 1973; Hennessy et al. 1976). Acquisition of appetitive response is not affected by corticosteroids (Bohus 1973; Hennessy et al. 1973) except when motivational level is low due to short deprivation (Bohus 1973).

Performance of various conditioned behaviors is usually improved by glucocorticoid administration. Dexamethasone improves the effectiveness of free-operant avoidance performance (Wertheim et al. 1967). Intertrial responsiveness is suppressed by cortisone treatment during active avoidance performance (Bohus and Lissak 1968). Dexamethasone improves differential reinforcement of low rate response (DRL) performance (Levine 1968). Improvement of reversal of a discriminative appetitive response follows corticosterone or dexamethasone treatment (Bohus 1973).

Retention of learned behavioral responses as studied in active and passive avoidance situations or with appetitive paradigms is profoundly influenced by glucocorticoids. Corticosteroids facilitate extinction of active avoidance responses (de Wied 1967; Bohus and Lissak 1968; Kovacs et al. 1976). Passive avoidance retention is suppressed by higher doses of corticosteroid (Bohus 1973, 1975; Kovacs et al. 1977). It is worth mentioning that low doses of corticosterone (100 μg/100 g) facilitated passive avoidance retention (Kovacs et al. 1977). Corticosteroids tend to facilitate extinction of conditioned taste aversion, but not consistently (Hennessy et al. 1977, 1980).

Recent observations suggest that very high doses of corticosteroids may also antagonize experimentally induced amnesia. Cortisol prevents the amnesic effect of cycloheximide (Nakajima 1975) and electroconvulsive shock (Nakajima 1978). Dexamethasone appears to counteract the amnesic effect of the protein synthesis inhibitor anisomycin (Flood et al. 1978). Corticosterone in a dose of 30 mg/kg, but not of 1.2 mg/kg is also antiamnesic (Squire et al. 1976).

2.3 Corticosteroid Effect on Behavior in Intact Rats: Intrinsic Effect on Brain or Suppression of ACTH Release?

Suppression of ACTH release by corticosteroids, in particular following repeated administration, is a classic feature of steroid effects. Behavioral changes induced by ACTH or related peptides are mostly opposite to those evoked by corticosteroids (see Bohus and de Wied 1980 for review). As we have seen, behavioral effects of adrenalectomy can also be ascribed to ACTH-like effects. Accordingly, steroid effects on behavior of intact rats may be pituitary-mediated. The majority of observations suggests an intrinsic behavioral activity of corticosteroids on brain mechanisms. The magnitude of effect of glucocorticoid treatment on active avoidance extinction (de Wied 1967) and on passive avoidance retention (Bohus 1973) does not correlate with the suppression of ACTH release. Both corticosterone and dexamethasone facilitate avoidance extinction in hypophysectomized rats (de Wied 1967). Implantation of glucocorticoids in the brain, in particular into the limbic-midbrain structures, does not influence pituitary ACTH release, but mimics the effect of systemic administration on avoidance extinction and passive avoidance retention (Bohus 1968, 1973). An exception is when cortisol is implanted in the median eminence. These implants effectively inhibit ACTH release and a modest facilitation of extinction occurs. The rate of extinction is correlated with the suppression of ACTH release (Bohus 1968). The synthetic steroid 6-dehydro-16-methylene-cortisol, which has been claimed as a practically pure ACTH-suppressor without gluco- and mineralocorticoid activity (Berthold et al. 1970), is clearly different from other steroids. Behavioral effect of this steroid upon systemic and intracerebral administration is highly correlated with the suppression of ACTH release (Bohus 1973). Accordingly, the majority of steroids affect behavior of intact rats through a direct interaction with brain mechanisms and it is exceptional that the inhibition of pituitary ACTH release can be considered as the cause of behavioral effects.

2.4 Localization of Behavioral Action of Corticosteroids in the Intact Rat

The limbic-midbrain system appears to be the site of action of corticosteroids on behavior of intact rats. This conclusion stems from studies using intracerebral implantation of crystalline corticosteroids. Implantation of a few μg of cortisol in the mesencephalic reticular formation, in the CA_1 and CA_2 layers of the dorsal hippocampus, the medial septal nuclei, anterior hypothalamus, the amygdaloid complex, or the posterior thalamus facilitates avoidance extinction in the intact rat as systemic cortisol treatment does. Implants in other hypothalamic or thalamic areas, the cerebral cortex and the ventral hippocampus are ineffective (Bohus 1968, 1970a). Corticosterone, dexamethasone, and progesterone implants in the thalamic parafascicular area also facilitate extinction (van Wimersma Greidanus and de Wied 1969; van Wimersma Greidanus et al. 1973). Similar effect of cortisol implants in the septal nuclei, the preoptic region, and the antero-lateral hypothalamus has been reported by Endröczi (1972) in adrenalectomized rats. Subsequently, attenuation of passive avoidance behavior by corticosterone, cortisol, and cortisone implants in those limbic-midbrain areas in which active avoidance extinction is affected has been observed (Bohus 1971, 1973, 1975). An interesting difference has, however, been noted in the effect of hippocampal corticosterone and cortisol implants. While cortisol implantation is followed by the attenuation of passive avoidance retention both shortly after learning and after a 1-day delay, corticosterone-implanted animals displayed a retention deficit only at the delayed test (Bohus 1971, 1975).

These observations suggest that steroid effects on avoidance behaviors of the intact rats involve multiple corticoid sensitive sites within the limbic-midbrain system. Since this system was regarded as a functional entity (Papez 1937), it seems that the whole system senses the presence of excess amounts of corticosteroids. The importance of hippocampal localization has been primarily emphasized on the basis of complex organizing function of this structure in behavioral adaptation (Bohus 1975).

2.5 Relationship Between Behavioral and Other Biological Activities of Steroids in the Intact Rat

Observations with natural and synthetic corticoids suggest comparative behavior activities of glucocorticoids such as corticosterone, cortisol, cortisone, and dexamethasone in intact rats (de Wied 1967; Bohus and Lissak 1968; Bohus 1971, 1973). Mineralocorticoids like aldosterone or deoxycorticosterone do not influence avoidance behaviors (de Wied 1967; Bohus and Lissak 1968; Bohus 1973). Accordingly, effects on avoidance acquisition and retention are primarily glucocorticoid rather than mineralocorticoid-bound effects. Gray (1976) on the other hand has found that enhancement of operant avoidance behavior by prestimulation is a mineralocorticoid-dependent phenomenon. In the intact rat, the physiological glucocorticoid of the rat, corticosterone, is the most effective in suppressing passive avoidance behavior (Bohus 1973). In contrast, van Wimersma Greidanus (1970) has found that dexamethasone, progesterone, pregnenolone, and 19-norprogesterone are as potent as corticosterone in facilitating active avoidance extinction. A sedative effect of large doses progesterone and related steroids (Gyermek et al. 1967) may explain the effectivity of progesterone on extinction. van Wimersma Greidanus et al. (1973) has excluded this possibility by showing that doses of progesterone, which affect extinction, do not suppress exploratory activity. Accordingly, a firm distinction between gluco- and mineralocorticoid effects in the intact rat is well established, but behavioral changes as observed following the administration of other pregnene-type steroids are not easy to explain neither on biological activity nor structural basis.

3 The Rat Brain Corticosterone Receptor System

Steroid hormones may act on cell metabolism via interaction with intracellular receptor proteins. The steroids form a complex with the receptor, which enters the cell nuclear compartment. The steroid-receptor complex binds to chromatin, where it triggers cellular responses involving modulation of genomic expression (RNA polymerase, RNA and protein synthesis). This is in short the mechanism of steroid action thought to take place in responsive tissues. Thus, if the steroid effect on behavior as described in Sect. 2 is mediated by such receptor sites for adrenocortical hormones, these actions should fulfill the same criteria for localization and specificity. We report here ongoing receptor studies with emphasis on heterogeneity of brain and pituitary binding systems for adrenocortical hormones and we discuss the findings in the light of steroid effects on brain function. Two aspects of steroid-cell interaction are considered: (1) Cell nuclear localization of adrenal steroids in brain and pituitary (in vivo studies) and (2) Binding specificity of the soluble adrenal steroid receptor system(s) in rat hippocampus and pituitary (in vitro studies). Most studies are performed with adrenalectomized animals to avoid the influence of variable levels of endogenous adrenocortical secretion products. As representative for the class of adrenal steroids we have taken data on corticosterone (B), deoxycorticosterone (Doc), and aldosterone (Aldo). Cortisol (F) is taken as an adrenal secretion product predominant in other species, but not in rat, and dexamethasone (Dex) as an example of an extremely potent synthetic glucocorticoid in peripheral tissues. Also progesterone (Prog) has been used.

3.1 Cell Nuclear Localization of Adrenal Steroids in Rat Brain and Pituitary

When a tracer dose of tritiated corticosterone is administered to adrenalectomized animals, the hormone is retained by cell nuclei of limbic brain regions, e.g., hippocampus, dorsolateral septum, and cortical and amygdaloid areas, but not in the hypothalamus or brain stem (McEwen et al. 1969, 1970). The principal localization is in neurons as shown by autoradiography (Gerlach and McEwen 1972; Stumpf and Sar 1975; Warembourg 1975; Rhees et al. 1975). Subsequent studies have shown that retention of adrenal glucocorticoids by limbic neuronal receptors is common for mammals, whether the principal hormone is corticosterone or cortisol (Gerlach et al. 1976). In intact rats corticosterone localizes with an essentially similar preference in extrahypothalamic structures as was shown by concentration measurement with a sensitive radioimmunoassay (McEwen et al. 1980). This is an important observation, since it confirms that the corticosterone receptor system operates as well in the presence of the whole spectrum of adrenocortical steroid secretion.

Among the five steroids tested in the dose range of about 10 nmol i.v., aldosterone is the only other steroid which shows a significant localization in hippocampal neurons and a similar regional difference of uptake in the brain (Ermisch and Rühl 1978; M. Moguilevsky, pers. communication, 1980; Veldhuis et al., unpublished results). What is different with corticosterone, however, is the time course and extent of cell nuclear labeling in adrenalectomized rats. Aldosterone uptake reaches maximum levels within 15 min, which is still half the amount of corticosterone attained in a 1-h interval (M. Moguilevsky, pers. communication, 1980; Veldhuis et al., unpublished results). Cortisol also shows a similar regional difference in uptake to corticosterone, but the amount recovered from hippocampal cell nuclei is only about 5% of that of corticosterone. Deoxycorticosterone uptake is even lower and progesterone uptake is negligible (Stumpf and Sar 1975; McEwen et al. 1976). It should be taken into account that in this study there is nothing that precludes sufficient steroid uptake when the dose of steroid is increased or when steady-state levels of the steroid are maintained in the plasma. Rather it seems that part of the differences observed in steroid uptake, with the possible exception of progesterone, are due to affinity toward plasma transcortin, metabolism and rate of penetration into brain cells. For different steroids there is large variation in the permeability of the blood-brain barrier (Pardridge and Mietus 1979). Differences in permeability may certainly contribute to the above findings, where a bolus injection is employed to measure steroid uptake. For instance, the binding of corticosterone to transcortin in contrast to aldosterone, which does not show any affinity to this plasma protein (Westphal 1971), may result in a more prolonged exposure of the brain cells to corticosterone and consequently a more extensive cell nuclear localization. Metabolism may have reduced the extent of cell nuclear localization of ^{3}H-Doc since this steroid undergoes metabolic transformation in brain tissue (Kraulis et al. 1975).

That properties intrinsic to the adrenal steroid receptor system determine the extent of cell nuclear localization also appears from uptake studies with dexamethasone. This synthetic steroid has a metabolic half-life in rat blood of about 4 h (de Kloet et al. 1974) and shows no affinity to plasma transcortin. Thus, if dexamethasone had a high affinity toward neuronal uptake and cell nuclear retention mechanism in the hippocampus, the long exposure to this steroid would result in a substantial accumulation. Yet, in the brain, dexamethasone shows no regional difference in cell nuclear uptake comparable to that of corticosterone (de Kloet et al. 1975). Dexamethasone uptake in hippocampal cell nuclei reaches a level of 10% of corticosterone; autoradiography has revealed that accumulation occurs predominantly in glial cells, circumventricular organs, vascular endothelial cells of choroid plexus, and neurons in the medial basal hypothalamus (Rees et al. 1975; Rhees et al. 1975; Warembourg 1975). The contrast in magnitude of cell nuclear uptake is not only restricted to the brain. Cell nuclei of the anterior pituitary gland and in particular those of immunoreactive ACTH cells show a great preference in uptake of the synthetic glucocorticoid (de Kloet et al. 1975; Rees et al. 1975, 1977).

The specificity of the cell nuclear uptake process of corticosterone in hippocampal neurons has been further defined in attempts to block the uptake with prior administration of excess competing steroid. When steroids (30 μg/100 g rat) are administered 30 min prior to administration of a tracer dose of ^{3}H-corticosterone, the cell nuclear uptake of the tracer is significantly suppressed in the relative potency of B = Aldo > Doc > F and no effect is observed of dexamethasone and progesterone pretreatment (Table 1). A tenfold increase in the dose of the two latter steroids also results in significant reduction of ^{3}H-B uptake. Some further insight into a possible mechanism of competition in the corticosterone uptake process can be obtained when the data are expressed as a ratio of cell nuclear to tissue uptake. This ratio (N/WH, Table 1) is reduced with prior administration of corticosterone, aldosterone, Doc, and cortisol, but is not affected by dexamethasone and progesterone. It seems likely, therefore, that corticosterone, aldosterone, Doc, and cortisol, at least in part, can interact with the same receptor system by reducing cell nuclear uptake of ^{3}H-B relatively more than tissue uptake of the tracer. Of these three steroids the Doc entrance in the cell nuclear compartment is severely hampered in vivo. Dexamethasone and progesterone suppress, in higher doses, cell nuclear and tissue uptake equally well, which certainly does not exclude that the two steroids have affinity toward the neuronal corticosterone receptor system. However, it is more likely that a different cellular retention mechanism exists in the brain for these steroids. Dexamethasone has a different cellular localization, is a poor blocker of corticosterone uptake, and the extent of cell nuclear uptake is not significantly influenced whether the rats are adrenalectomized or not (de Kloet et al. 1977b). Also progesterone binds to different receptor sites. One aspect in the cell nuclear uptake process deserves consideration. The large differences observed between corticosterone and steroids such as dexamethasone, Doc, cortisol, and progesterone disappear when cell nuclear uptake is studied in vitro in tissue slices (de Kloet et al. 1975; McEwen et al. 1976). Although, ^{3}H-B uptake at saturating steroid concentration (2×10^{-8} M) is still highest, that of the other four steroids has increased considerably. This observation may stress the significance of metabolism, transcortin binding, and rate of penetration, for the very pronounced uptake of corticosterone in vivo. It also

Table 1. Blockade of hippocampal cell nuclear uptake of ^{3}H-corticosterone in vivo with prior administration of unlabeled steroids.

Pretreatment [a]	Cell nuclear retention [b] (fmol/mg protein)		Uptake cell nuclei/Uptake tissue (N/WH)	
	Mean ± SEM	No. of observations	Mean ± SEM	No. of observations
Saline	123 ± 7.7	14	4.1 ± 0.4	14
30 µg dose				
B	23 ± 5	3	2.5 ± 1.0	3
Aldo	20 ± 5	3	1.5 ± 0.8	3
Doc	47 ± 15	3	3.1 ± 1.0	3
F	84 ± 8	3	2.7 ± 0.7	3
Prog	97 ± 12	3	4.1 ± 1.2	3
Dex	103 ± 13	3	5.2 ± 2.3	3
300 µg dose				
B	3.3 ± 1.4	3	0.6 ± 0.2	3
Doc	11 ± 4.0	3	1.2 ± 0.9	3
F	36 ± 2.5	3	3.2 ± 0.2	3
Dex	50 ± 20	2	4.3 ± 1.1	2
Prog	83 ± 30	2	5.0 ± 0.4	2

[a] Steroids were administered in a dose of 30 µg/100 g rat or 300 µg/100 g rat 30 min prior to administration of a tracer dose of ^{3}H-corticosterone (50 µ Ci) to rats adrenalectomized 3 days previously.

[b] One hour after the injection of the tracer dose ^{3}H-B the animals were killed by decapitation, the hippocampus was dissected, and the purified cell nuclear fraction was isolated. ^{3}H-corticosterone was extracted from this fraction and counted in a liquid scintillation counter.

may be explained by the so-called cryptic receptors discovered recently by Turner and McEwen (1980). The existence of "cryptic" receptors became apparent when it was shown that doses of corticosterone, which saturate the cell nuclear uptake process in vivo in the hippocampus, do not occupy more than 40% of the maximal available binding sites of the receptor system as determined in vitro in cytosol. These sites, detected in vitro, may not operate in vivo and according to this view in vivo cell nuclear uptake is the best criterium for the biochemical end point in the adrenal steroid-receptor interaction.

3.2 Binding Specificity of Soluble Adrenal Steroid Receptor Sites in Rat Hippocampus: In Vitro Studies

Scatchard analysis of ^{3}H-B-binding to cytosol receptor sites has revealed a linear plot. This analysis suggests that the two components, which have been separated by isoelectric focusing (MacLusky et al. 1977) or DEAE-cellulose anion exchange chromatography (de Kloet and McEwen 1976) bind corticosterone with only small differences in affinity. Relative binding affinity to the sites labeled with ^{3}H-B is in the order of B = Doc > Prog = Dex > Aldo (Table 2). Binding of ^{3}H-Dex also gives a linear Scatchard plot. Relative binding affinity of the steroids to the dexamethasone-labeled sites is similar to that of corticosterone. The order is now: Doc = B > Prog > Dex > Aldo (Table 2). In contrast, using ^{3}H-aldosterone as the binding ligand did not provide a linear Scatchard plot. Two lines could be constructed. This curvilinear pattern is indicative for heterogeneity or can be explained by negative cooperative interactions between a homogenous class of receptor sites at high ligand concentrations. Inclusion of a onefold excess of unlabeled corticosterone linearizes the Scatchard plot and leaves that population of binding sites with highest affinity for aldosterone. Thus, corticosterone competes for sites, which have a lower affinity for ^{3}H-Aldo and which presumably represent the high affinity corticosterone binding sites. When the competing corticosterone concentration is further increased to tenfold, then binding of ^{3}H-Aldo to the high affinity sites is also decreased.

Table 2. Relative binding affinities (RBA) of various steroids for ^{3}H-corticosterone- and ^{3}H-dexamethasone-labeled binding sites in hippocampal cytosol.

Steroid	RBA [a] for ^{3}H-B binding site	RBA [a] for ^{3}H-Dex binding site
B	100	198
Aldo	15	44
Doc	96	248
Prog	42	134
Dex	33	100

[a] The concentration of each competitor (c. IC_{50}) required to decrease bound radioactivity of the ligand by 50% was determined and compared to the concentration of unlabeled ligand (1. IC_{50}) required to obtain the same displacement. The ratio (1. IC_{50}/c. IC_{50}) × 100 is the relative binding affinity (RBA).
The RBA of corticosterone and of dexamethasone were taken as 100.

All these in vitro binding studies were performed with cytosol, which was first prepared and then used for the binding experiments. Absence of steroid renders the receptor sites more susceptible to enzymatic inactivation and this process starts immediately after cell disruption at homogenization at 0°C. In fact, the time course of inactivation has been used to distinguish between corticosterone- and dexamethasone-labeled sites in brain cytosol, since the binding sites for dexamethasone were more rapidly inactivated (de Kloet et al. 1975).

3.3   Three Populations of Adrenal Steroid Receptor Sites in Rat Brain

At present we have arrived at the conclusion that there are three populations of adrenal steroid receptor sites in rat brain. This inference is based on autoradiographical and biochemical studies of steroid uptake, on binding studies of steroid-receptor interaction in vitro and on analytical techniques such as isoelectric focusing and ion exchange chromatography of the steroid-receptor complexes.

The first distinction is that between gluco- and mineralocorticoid sites (Moguilevsky and Raynaud 1980; Anderson and Fanestil 1976), which may even coexist in the same neurons (Ermisch and Rühl 1978). The second distinction refers to heterogeneity within the class of glucocorticoid receptor sites. One population is the neuronal corticosterone receptor system, with highest preference for the principal endogenous glucocorticoid of the rat. The other population shows resemblance to the glucocorticoid receptor, which is commonly observed in peripheral target tissues including the anterior pituitary (Feldman et al. 1978). These receptors prevail in glial cells as shown in studies with enucleated rat optic nerve (Meyer et al. 1978) and glial tumor cell line C_6, which derive from rat brain (de Vellis et al. 1974; McGinnis and de Vellis 1978).

Intracellular processing or proteolysis could account for the heterogeneity observed in glucocorticoid receptors (Wrange 1979; Sherman et al. 1978; Agarwal 1978). If so, then compartmentalization of glucocorticoid receptor sites in the brain in different cell types, e.g., neurons and glial cells, still would represent functionally different entities. Ultimately, heterogeneity of the receptors should be proven by purification of steroid-receptor complexes and comparison with the original tissue localized receptor complexes and comparison with the original tissue localized receptor proteins. Some progress has been made in purification of a single population of corticosterone receptor proteins (de Kloet and Burbach 1978).

Receptor heterogeneity reminiscent to that in the brain has been observed in the rat kidney (Funder et al. 1973a, b; Strum et al. 1975). Based on binding kinetics these sites have been termed arbitrarily as mineralocorticoid receptors (type I), glucocorticoid receptors (type II) with highest affinity for synthetic glucocorticoids, and corticosterone specific receptor sites (type III). Mineralocorticoid receptors have been found besides the renal cortex in a number of tissues of primarily epithelial origin (Marver and Schwartz 1980; Funder et al. 1973b). Type II, glucocorticoid receptors occur in almost every tissue, consonant with the widespread effects of the hormones (Feldman et al. 1978; Ballard and Ballard 1974; Baxter and Rousseau 1979). Type III receptor have so far only been demonstrated in the kidney-collecting tubules (Strum et al. 1975) and in neurons. Crossover interaction of the respective steroids among the three populations of receptor sites is likely to occur. The extent of crossover depends on the affinity of the steroid for the particular receptor population, the factors mentioned earlier, and the circulating hormone level.

3.4 Intracellular Transcortin-like Molecules

Apart from the three populations of adrenal steroid receptor sites now recognized in brain tissue, there is also firm evidence for a fourth population of tissue-localized, high-affinity binding sites, which is selective for corticosterone. These binding sites are found in substantial amounts in pituitary cytosol and resemble plasma transcortin in molecular weight, isoelectric focusing profile, salt precipitability, immunological properties, and lack of affinity to DNA (Koch et al. 1976, 1978; de Kloet and McEwen 1976; MacLusky et al. 1977). The transcortin-like binding system (TL) is not due to blood contamination since it persists after extensive perfusion of the rats when killed and can be extracted from isolated pituitary cells (de Kloet et al. 1977a, b; Koch et al. 1977). TL molecules remain associated with plasma membranes (Koch et al. 1978) and preliminary immunocytochemical observations suggest an intracellular localization (de Kloet and Voorhuis, unpublished results). The origin of the TL system may well be the blood since changes induced in plasma transcortin level by adrenalectomy, estrogen, or thyroxin treatment are reflected in the pituitary (Koch et al. 1978; de Kloet et al. 1977a). The TL system is absent in the pituitary 1 week after birth, when plasma transcortin is undetectable (Sakly and Koch 1981).

There is evidence that the TL-binding system, associated with plasma membranes, is actively involved in the rate of uptake and intracellular distribution of corticosterone in the pituitary gland (Koch et al., personal communication). That membrane components may be involved in the uptake process has been reported previously for estradiol binding to uterine cells (Milgrom et al. 1973) or to liver cells (Pietras and Szego 1977) and for glucocorticoid binding to AtT-20 tumor cell line of pituitary (Harrison et al. 1977). Whether corticosterone diffuses into the cell, as is commonly thought, or enters as complex with plasma transcortin is not known. The ultimate rate and extent of cell nuclear uptake of corticosterone is, however, much lower than that of dexamethasone in vivo and in vitro in tissue slices (de Kloet et al. 1975) and pituitary tumor cells (Svec and Harrison 1979). This difference is not only due to the affinity of a steroid to the receptor and its intrinsic ability to promote cell nuclear translocation, but also seems a consequence of the presence of the intracellular TL system. The intracellular TL system is thought to compete with the receptor. Dexamethasone bypasses this mechanism since it does not bind to transcortin (de Kloet et al. 1977a, b; Koch et al. 1977).

These aspects of glucocorticoid-cell interaction in the pituitary may help to explain the extremely potent suppressive effect of dexamethasone on stress-induced pituitary ACTH release (de Kloet et al. 1974) and also why under such conditions only supraphysiological corticosterone levels are effective (Smelik and Papaikonomou 1973). The dexamethasone effect requires RNA synthesis (Nakanishi et al. 1977) and may represent the delayed feedback mechanism (Dallman and Yates 1969). The TL-binding system is virtually undetectable in the rat hippocampus, but significant amounts can be extracted from the hypothalamus (de Kloet et al. 1977a, b; MacLusky et al. 1977). Accordingly, it is likely that the pituitary and the hypothalamus are not susceptibe to small changes in blood corticosterone concentration, in contrast to the hippocampus.

3.5 Evaluation of Uptake and Binding Studies: Discrepancy with Early Behavioral Observations

The behavioral studies in the intact rat have shown that all steroids with glucocorticoid activity, but also progesterone and pregnenolone, may have similar effects. The doses used were mainly in the mg range; the steroids were sometimes equipotent. The steroid-responsive sites were found widely distributed over limbic-midbrain regions. However, a dichotomy was observed, for instance, upon implantation of the glucocorticoids in the hippocampus and medial basal hypothalamus; hippocampal implants affected conditioned behavior, but not the pituitary ACTH release, while hypothalamic implants did the opposite.

These behavioral observations are in striking contrast to the properties of the corticosterone receptor system in rat brain, which is thought to mediate steroid action on nerve cells. The neuronal corticosterone receptor has a stringent specificity, which is expressed most clearly in cell nuclear retention. The corticosterone receptor system operates within the limits of physiological plasma corticosterone levels (McEwen et al. 1974; Grosser et al. 1973) with half maximal saturation at plasma levels of 5 μg%, which are morning levels under resting conditions.

4 Specific Behavioral Effects of Corticosterone in the Rat

The reviewed behavioral effects of steroids and the characteristics of the brain corticosteroid receptor systems are in conflict in two cardinal points. Firstly, steroid effects on behavior do not show the strict structural specificity which is required for an interaction with the limbic corticosterone receptor system in the rat. Secondly, the low capacity of the corticosterone receptor system is not consistent with the possibility that supraphysiological doses of adrenal steroids affect behavior through these receptors in intact rats. Accordingly, serious doubts may be raised concerning a specific role of adrenal steroids, in particular corticosterone, in behavioral adaptation. In addition, observations in adrenalectomized rats long after surgery suggest that the behavioral alterations may be considered as a consequence of ACTH-related peptide effects rather than the absence of corticosteroids. Furthermore, short-term behavioral impairments are due to adrenomedullary dysfunction.

In order to resolve the apparent contradictions concerning the physiological importance of corticosterone in behavioral adaptation, we decided to reinvestigate the influence of adrenalectomy on rat behavior. The strategy of these experiments was based on the classical endocrine approach, i.e., removal of the source of the adrenal steroids and replacement with physiological amounts of endogenously occurring hormones. In addition, in the experimental design it was attempted to eliminate ACTH-dependent behavioral alterations which may follow adrenalectomy. Therefore, short intervals (1 or 2 h) between adrenalectomy and the behavioral manipulation were selected. The behavioral paradigm, used for the majority of the experiments, was a one-trial learning, passive (inhibitory) avoidance test as described by Ader et al. (1972). Briefly, rats, like other rodents, prefer dark to light. If placed on an extensively lit, elevated platform, they readily enter a dark chamber to which the platform is attached. Following adaptation to the apparatus and pretraining to enter the dark, the rats receive an unescapable electric footshock (0.5 mA, a.c. of 1 s duration) through the grid floor of the dark compartment. Passive (inhibitory) avoidance behavior manifests itself in a long latency to reenter the dark compartment from the elevated platform 24 h after the learning trial (retention test). Confinement of rats for 5 min shortly (3 h) after the learning trial to the dark compartment where they have experienced punishment (so-called forced extinction training) results in a complete extinction of the passive avoidance response (Robustelli et al. 1972; Bohus 1974). The extinction paradigm was selected because of former experience that this type of test is the most sensitive for demonstrating hormonal modulation of behavior (Bohus and de Wied 1980).

4.1 Adrenalectomy-Induced Impairment of Forced Extinction Behavior: Normalization with Corticosterone

In order to investigate whether the adrenals are essential for normal extinction behavior, adrenalectomy was performed under light ether anesthesia at different phases of the behavioral training. As shown in Table 3 adrenalectomy before the forced extinction impaired extinction behavior as studied by a retention test 24 h later. The adrenalectomized rats, in contrast to sham-operated animals, show passive avoidance behavior as they have not been subjected to forced extinction. Removal of the adrenals immediately after the forced extinction training or 1 h before the retention test fails to affect extinction behavior (Bohus 1974).

That the behavioral deficit of the adrenalectomized rats was due to the absence of sufficient amount of corticosterone during the forced extinction training is suggested by the following observations.

Table 3. Effect of adrenalectomy (ADX) or sham adrenalectomy (SAX) on the retention of a passive avoidance response following forced extinction: dependence on the time of surgery.

Surgery	Time of surgery	Avoidance latency [a]	
		Forced extinction	No forced extinction
ADX	1 h prior to	86.5 (12) [b]	84.0 (12)
SAX	forced extinction	6.0 (11)	131.5 (10)
ADX	Immediately after	8.5 (6)	160.5 (6)
SAX	forced extinction	15.5 (6)	120.0 (6)
ADX	1 h prior to	14.5 (6)	115.5 (6)
SAX	retention test	7.5 (6)	148.0 (5)

( ), No. of observations.
[a] Medium in s at the retention test.
[b] $P < 0.02$ vs sham-operated controls (Mann-Whitney test).

Corticosterone substitution of rats adrenalectomized before the forced extinction training normalizes extinction behavior, i.e., very short avoidance latencies at the retention test, if the treatment is given immediately after surgery (Table 4). The administration of the steroid before the retention test fails to amend the impaired behavior of the adrenalectomized rats. If the treatment is given both before the extinction training and the retention test, normal extinction behavior occurs. This observation excludes the possibility that the behavior of the adrenalectomized and substituted animals and of the operated controls is different because of different hormonal states during extinction training and the retention test. Pappas and Gray (1971) showed that dexamethasone may function as a cue for the behavior of the animals. They observed that intact animals receiving dexamethasone both before training and test of an avoidance response were not different from controls while dexamethasone given either before training or test suppressed avoidance behavior. A similar kind of "state dependency" was suggested by Stewart et al. (1967), using progesterone treatment.

Table 4. Effect of corticosterone substitution on the retention of a passive avoidance response in rats adrenalectomized before the forced extinction training.

Time of treatment	Treatment	Avoidance latency [a]	
60 min before forced extinction	Corticosterone [b]	9.0 [c]	(9)
	Saline	60.0	(9)
60 min before retention test	Corticosterone	87.0	(9)
	Saline	165.0	(9)
Both before forced extinction and retention	Corticosterone	21.0 [c]	(6)
	Saline	95.0	(6)

( ), No. of observations.

[a] Median in s at the retention test (24 h forced extinction training).

[b] 300 μg/100 g body wt. subcutaneously.

[c] $P < 0.01$ (Mann-Whitney test).

4.2 Specificity of Corticosterone in Correcting Extinction Behavior of Adrenalectomized Rats

The above observations suggest that the presence of an intact adrenal cortex or sufficient amounts of exogenous corticosterone during forced extinction training is an essential condition for extinction behavior of the rat. If the hippocampal neuronal glucocorticoid receptor system is involved in this behavior, a highly specific effect of corticosterone in adrenalectomized rats may be expected. Accordingly, the influence of treatment with the naturally occurring adrenal steroids such as progesterone and deoxycorticosterone and with the synthetic dexamethasone on the extinction behavior of adrenalectomized rats has been subsequently investigated (Bohus and de Kloet 1977, 1981). These experiments, as shown in Table 5, indicate a highly specific effect of corticosterone in normalizing extinction behavior of adrenalectomized rats. Significantly shorter avoidance latencies which indicate a normal extinction behavior were observed only in rats which received corticosterone. Neither progesterone, nor Doc, nor dexamethasone show any agonistic activity up to the dose level of 300 μg/ 100 g body wt. Consonant with the notion of highly specific effects of corticosterone on brain function are the recent observations of Micco et al. (1979, 1980). They found that corticosterone but not dexamethasone influenced extinction of an appetitive response in rats adrenalectomized 4 weeks previously (Micco et al. 1979). Furthermore, Micco and McEwen (1980) showed that corticosterone but not dexamethasone normalizes extinction behavior of rats adrenalectomized 2 days prior to the extinction training of an appetitive task.

Table 5. Comparative effects of various steroids on the impaired extinction behavior of adrenalectomized rats.

Treatment [a]	Dose [b]	Avoidance latency [c]	
Saline	–	86.5	(12)
Corticosterone	30	24.0 [d]	(12)
	300	9.0 [e]	(11)
Progesterone	30	88.0	(8)
	300	99.5	(8)
Deoxycorticosterone	30	72.5	(6)
	300	88.0	(6)
Dexamethasone	30	87.0	(12)
	300	69.0	(12)

( ), No. of observations.

[a] Treatment was given immediately after adrenalectomy.
[b] μg/100 g body wt. subcutaneously.
[c] Median in s at the retention test.
[d] $P < 0.05$ vs saline-treated controls (Mann-Whitney test).
[e] $P < 0.01$ vs saline-treated controls (Mann-Whitney test).

As previously mentioned, aldosterone and Doc may interact, at least partially, with the same neuronal receptor system in the brain as corticosterone. Although Doc failed to show agonistic behavioral activity, aldosterone seemed to be a better ligand for further investigating the specificity of hippocampal corticosterone receptor system in forced extinction behavior. In contrast to Doc, aldosterone is taken up by the hippocampal neuronal cell nuclei and this steroid is also a better competitor for corticosterone uptake than Doc. Aldosterone in doses of 30 and 300 μg/100 g body wt. fails to mimic the effect of corticosterone on forced extinction behavior of rats adrenalectomized and replaced by steroids 1 h prior to the extinction training.

Absence of an effect of aldosterone may be explained by a low affinity of this steroid for corticosterone receptor sites in the hippocampus. Interestingly, aldosterone and also Doc failed to affect avoidance behavior in intact rats while all steroids with glucocorticoid activities did (de Wied 1967; Bohus 1973). It seems therefore that the nature of neural processes induced by corticosterone and aldosterone further accentuate the specificity of steroid-hippocampal interaction in behavioral regulation. The observation of Gray (1976) is in favour of this notion. He showed that prestimulation-induced enhancement of an avoidance response is absent in adrenalectomized rats. Aldosterone and Doc, but not corticosterone, correct this behavioral deficit.

4.3 Ineffectiveness of Dexamethasone on the Extinction Behavior of Adrenalectomized Rats

The above-mentioned observations, showing a highly specific effect of corticosterone on the extinction behavior of adrenalectomized rats, are consonant with the strict specificity of the hippocampal neuronal receptor system. In order to exclude the possibility that time-, dose-, or uptake-related differences are responsible for the specific effect of corticosterone versus dexamethasone, additonal experiments have been performed in the adrenalectomized rats. In these studies the adrenals were removed 2 h prior to the forced extinction training (Sokolova et al., in preparation).

Corticosterone substitution normalizes the extinction behavior of adrenalectomized animals whether the treatment was given 120, 90, 60, or 30 min prior to the forced extinction training. Accordingly, sufficient amounts of this steroid were present in the brain even after a long (120 min) or short (30 min) interval between treatment and training. Dexamethasone substitution at all intervals failed to affect the behavior of adrenalectomized rats. Therefore, the specificity of corticosterone and the ineffectiveness of dexamethasone are unrelated to the treatment-extinction training interval.

Administration of dexamethasone 60 min before the training up to a dose of 15 times higher than an already effective corticosterone dose fails to normalize the behavior of adrenalectomized rats. Accordingly, the differences between corticosterone and dexamethasone are not related to the dose of the latter steroid.

Administration of corticosterone into a lateral cerebral ventricle is also effective in normalizing the extinction behavior of adrenalectomized rats when 50 ng or more was given. Dexamethasone in a dose of 10 times higher than the minimally effective dose of corticosterone slightly increases the number of rats which displayed extinction behavior. A dose of 750 ng of dexamethasone is almost as effective as 50 ng of corticosterone. It is therefore likely that these very large amounts of dexamethasone, when applied in the vicinity of the hippocampus, may exert their effects via the corticosterone sensitive receptor mechanism.

4.4 Antagonism by Progesterone, Deoxycorticosterone, and Dexamethasone of the Effect of Corticosterone on Extinction Behavior

Additional evidence for the specificity of corticosterone in affecting extinction behavior of adrenalectomized rats has been obtained by showing that progesterone, deoxycorticosterone, and dexamethasone pretreatment abolish the influence of corticosterone substitution on extinction behavior (Bohus and de Kloet 1981). The strategy of behavioral experiments was based upon the assumption that occupation of hippocampal receptor sites by competing steroids would abolish the binding of corticosterone to cytosol receptors and its transport to the cell nuclei.

Table 6. Antagonistic properties of progesterone, deoxycorticosterone, and dexamethasone on the extinction behavior of adrenalectomized rats, substituted with corticosterone.

Pretreatment [a]	Treatment [b]	Dose [c]	Avoidance latency [d]	
Saline	Saline	–	84.5	(15)
Saline	Corticosterone	30	24.5	(8)
	Corticosterone	300	9.0	(9)
Corticosterone	Corticosterone	30	20.0	(6)
	Corticosterone	300	7.5	(6)
Progesterone	Corticosterone	30	75.5 [e]	(8)
	Corticosterone	300	60.0 [e]	(9)
Deoxycorticosterone	Corticosterone	30	100.0 [f]	(6)
	Corticosterone	300	12.5	(6)
Dexamethasone	Corticosterone	30	77.5 [e]	(8)
	Corticosterone	300	56.5 [e]	(10)

[a] Pretreatment was given 120 min before forced extinction in a dose of 30 μg/100 g body wt.

[b] Corticosterone substitution was given 60 min before forced extinction.

[c] In μg/100 g body wt.

[d] Median in s at the retention test.

[e] $P < 0.01$ vs saline- and corticosterone-treated rats (Mann-Whitney test).

[f] $P < 0.05$

As shown in Table 6, pretreatment with a low dose of progesterone and dexamethasone is able to block the behavioral effect of equimolar or 10 times higher doses of corticosterone. Deoxycorticosterone pretreatment abolishes the effect of 30 μg/100 g body wt. of corticosterone, but not of that of the higher dose. The observation with deoxycorticosterone can be explained by our recent in vivo observations on the antagonistic properties of this steroid of the hippocampal corticosterone uptake in cell nuclei (see previous section). The antagonistic properties of dexamethasone and progesterone, although agreeable to the competition studies in vitro (McEwen et al. 1976), are difficult to reconcile with the finding that these steroids are practically unable to block the translocation of corticosterone into the hippocampal cell nuclei in vivo (see Sect. 5). Moreover, our observations suggest that the presence or continuous "renewal" of the corticosterone-receptor complex in the cell nuclei is essential for normal extinction behavior and the action of corticosterone is to maintain certain neural processes which are essential for an effective forced extinction.

4.5 Corticosteroids and the Exploratory Behavior of Adrenalectomized Rats

The specificity of corticosterone for normalizing the extinction behavior of adrenalectomized rats in a fear-motivated behavioral test raises the question of whether this finding is to generalize across aversively and appetitively motivated extinction behaviors, and other adaptive behavioral responses. As far as the generalization of an effect on extinction behaviors is concerned, Micco et al. (1979) and Micco and McEwen (1980) showed that impaired extinction of a food-reinforced straight runway response in rats, adrenalectomized either several weeks before or 2 days prior to the extinction test, is normalized by corticosterone, but not by dexamethasone. Interaction with the hippocampal receptor system for corticosterone system is likely because of the involvement of hippocampus in extinction processes as a corollary of behavioral inhibition (Douglas 1967; Kimble 1968).

Another behavioral response which may depend on the integrity of the adrenal cortex is exploratory behavior in an open field. If a rat is placed in a novel environment, the primary behavioral response is to explore the field. The exploration is composed of an ambulatory (horizontal movements: running and sniffing) and a rearing component (vertical movements). These behavioral activities are interrupted by short periods of grooming, which may be considered as a displacement activity in the novel environment. Defecation and urination do occur as well, and are regarded as an index of emotionality. The number of floor units crossed in the circular field (ambulation) and of rearings, and the frequency and duration of grooming episodes serve as the measures of exploratory behavior (Weijnen and Slangen 1970).

Although adrenalectomy fails to influence the exploratory activity in the open field during a short (2 min) observation session (Paul and Havlena 1962; Joffe et al. 1972), adrenalectomized male rats are less active than sham-operated controls in longer (5 min) observation sessions (McIntyre 1976; van Deusen and Bohus, unpublished). It seems therefore, that there is a more rapid decline in exploratory activity of adrenalectomized rats with the extension of observation period. Although these observations were performed 10 or more days after adrenal removal, it is unlikely that ACTH is reponsible for this behavioral phenomenon. Neither ACTH (Moyer 1966) nor fragments such as ACTH 4–10 (Bohus and de Wied 1966; Weijnen and Slangen 1970) influenced exploratory behavior of intact rats.

Recently we have investigated the development of changes in exploratory behavior of rats following adrenalectomy. Male rats were adrenalectomized under short ether anesthesia and subjected to a 5-min open field test, 1, 4, 24, 72, or 240 h after adrenalectomy (Bohus et al., unpublished).

Ambulation is not changed by adrenalectomy up to 72 h after surgery. However, 240-h adrenalectomized rats ambulate significantly less (84.6 ± 5.2) than the sham-operated controls ($103.6 \pm 5.2; P < 0.02$). Rearing activity is significantly suppressed 1 h after adrenalectomy (8.6 ± 1.4 vs $16.8 \pm 2.3; P < 0.01$), but no differences occur 4 or 24 h after surgery. This behavioral activitiy is again slightly decreased 72 h after adrenal removal and a significant suppression is seen 240 h following the operation (9.8 ± 2.5 vs $15.3 \pm 1.7; P < 0.05$). Grooming behavior and defecation scores were not affected by adrenalectomy.

Table 7. The effect of simultaneous administration of ACTH 4−10 and corticosterone or dexamethasone on exploratory behavior of adrenalectomized rats.

Treatments [a]		Ambulation	Rearing	
Intraventricular [b]	Peripheral [c]			
Saline	Saline	71.3 ± 5.5	12.4 ± 1.7	(9)
ACTH4−10	Saline	90.3 ± 6.3 [d]	16.7 ± 3.2	(10)
Saline	Corticosterone	99.6 ± 7.4 [e]	19.4 ± 3.0 [d]	(9)
ACTH 4−10	Corticosterone	83.5 ± 2.9	20.3 ± 2.1 [e]	(9)
Saline	Dexamethasone	78.5 ± 5.8	19.5 ± 3.1 [d]	(11)
ACTH 4−10	Dexamethasone	89.1 ± 5.0 [e]	12.0 ± 1.3	(12)

(   ) No. of rats.
[a] Treatments were given 1 h before the tests.
[b] ACTH 4−10 was given in a dose of 30 ng per rat.
[c] Corticosterone and dexamethasone were given in a dose of 30 μg/100 g body wt.
[d] $P < 0.05$ vs saline and saline controls (Mann-Whitney test).
[e] $P < 0.01$ vs saline and saline controls (Mann-Whitney test).

The effects of corticosteroid replacement of adrenalectomized rats by a single administration of corticosterone or dexamethasone before the open field test 240 h after surgery are shown in Table 7. Corticosterone, but not dexamethasone administration, results in an increase of ambulatory activity toward the sham-operated level. Rearing activity was increased by both corticosterone and dexamethasone.

These observations suggest that the ambulatory and rearing components of exploratory behavior require an intact adrenal cortex in the rat. The two behavioral components are, however, differently affected by the absence of the adrenals. Decreased ambulatory activity was not observed before day 10 postoperatively. Facilitation of ambulation by corticosterone toward normal levels indicate the involvement of the adrenal cortex. However, brain corticosterone receptors are already deprived of their endogenous ligands shortly after adrenal removal. An absence of an effect of dexamethasone on ambulation argues against the involvement of an increased pituitary ACTH release, which follows adrenalectomy. Dexamethasone in a dose used here effectively suppresses ACTH release. It is therefore more likely that decreased function of some neurotransmitter systems, which takes time to develop in the absence of corticosterone, is responsible for a decreased ambulatory activity.

The specificity of corticosterone to facilitate ambulatory behavior of adrenalectomized rats suggests the involvement of a corticosterone-specific brain receptor system. Similar specificity in facilitating rearing activity is, however, absent. Although observations do not favor an idea of different control of ambulatory and rearing activities (Heybach et al. 1978; Kovacs and Telegdy 1978), the development of decreased activities in the absence of adrenals is not entirely parallel. Accordingly, differences in the specificity of corticosteroids to affect different components of exploratory behavior of adrenalectomized rats remains an issue to be solved.

4.6 Interactions Between the Behavioral Effects of ACTH-Related Peptides and Corticosteroids

As mentioned earlier, the effects on adaptive behavior of corticosteroid and ACTH-related peptides are mainly of opposite character. That ACTH and corticosteroids may interact in affecting behavior has been suggested by studies with intracerebral steroid implantation and systemic injection of ACTH in the rat. Implantation of cortisone in the posterior thalamic parafascicular area abolishes the influence of ACTH on the extinction of an active avoidance response (Bohus 1970b). The parafascicular nucleus is involved in the mediation of the effect of ACTH-related peptides on avoidance extinction. Electrolytic lesion of this region prevents the effects of α-MSH (Bohus and de Wied 1967) and ACTH 4–10 (van Wimersma Greidanus et al. 1974) on avoidance behavior. In addition, local injection of ACTH 1–10 increases resistance to extinction (van Wimersma Greidanus and de Wied 1971), i.e., mimics the behavioral effect of systemically administered peptide (Bohus and de Wied 1966). Systemic administration of ACTH, however, counteracts the influence of cortisone implants in forebrain areas, such as the anterior hypothalamus, amygdala, and the medial septum – that is, it prevents the rapid extinction of an active avoidance response (Bohus 1970b). These observations suggest that cortisone is able to block the effect of ACTH at its site of action (parafascicular area), whereas the influence of ACTH is dominant over cortisone effect on forebrain structures.

Since cortisone is not a specific ligand for the brain corticosterone receptor system, it was of interest to reinvestigate the interactions between ACTH-related peptides and glucocorticoids in relation to behavior. Exploratory behavior in an open field has been selected as behavioral paradigm (Bohus, unpublished). Adrenalectomized rats display a decreased exploratory activity 10 days postoperatively and corticosterone, but not dexamethasone replacement, normalizes behavior.

ACTH 4–10, an ACTH-related neuropeptide, similarly to corticosterone, facilitates ambulatory activity upon intracerebroventricular administration (Table 7). In spite of the same direction of action on ambulation, simultaneous administration of the peptide and corticosterone fails to alter the exploratory behavior. Ambulatory activity of saline-treated controls is not different from that of ACTH 4–10 + corticosterone-treated rats, while the latter group displays significantly less ambulatory activity than either ACTH 4–10 or corticosterone-treated adrenalectomized rats. Administration of dexamethasone together with ACTH 4–10, on the other hand, does not abolish the peptide effect on ambulation.

ACTH 4—10, unlike corticosterone or dexamethasone, fails to influence rearing activity in the open field (Table 7). ACTH 4—10- and dexamethasone-treated rats display a rearing activity which was indistinguishable from that of the adrenalectomized rats. Simultaneous administration of ACTH 4—10 and corticosterone induces a similar increase in rearing as does corticosterone treatment alone.

These observations suggest the existence of multiple interactions between the behavioral effects of ACTH-related peptides and corticosteroids. Effects on ambulatory activity suggest the involvement of a corticosterone-specific receptor system, since dexamethasone failed to influence this component of exploratory behavior and an interaction with ACTH 4—10 was also absent. Corticosterone specificity of this interaction indirectly suggests the involvement of hippocampus, but the mode of interaction is not yet clear. Neuropeptides, like ACTH 4—10, may regulate corticosterone-binding capacity in the hippocampus upon chronic administration (see next section).

The interactions between ACTH 4—10 and corticosterone and dexamethasone in affecting rearing are rather complex. First of all, normalization of rearing in adrenalectomized rats is not corticosterone specific. Secondly, ACTH 4—10 itself fails to affect this behavioral component. Thirdly, ACTH 4—10 suppresses, but only dexamethasone-induced rearing. Accordingly, one may not exclude the possibility that dexamethasone but not corticosterone interacts with that mechanism which mediates the intrinsic behavioral activity of ACTH-related peptides.

5   Biochemical Correlates of Corticosterone Effects on Behavior

5.1   Receptor Studies

The steroid specificity of the effects on forced extinction and exploratory behavior indirectly suggests that such effects are mediated by the hippocampal corticosterone receptor system. Corticosterone is unique in displaying agonistic action on these behaviors and in promoting translocation of the receptor to the cell nucleus. Aldosterone, the only steroid resembling corticosterone in cell nuclear retention was ineffective on behavior. Since also the mineralocorticoid Doc was ineffective, these observations provide additional arguments for two functionally different populations of receptor sites (gluco- and mineralocorticoid) capable in cell nuclear translocation. However, there is a considerable cross specificity as seen from the effective blockade of corticosterone cell nuclear transport in vivo by Doc and aldosterone pretreatment. Therefore, these steroids most likely accomplish their antagonism on corticosterone by blocking cell nuclear translocation of B (Fig. 1). Dexamethasone and progesterone probably exert their antagonistic action via competition at the receptor level. In accordance with the observation of McEwen and Wallach (1973) we have observed that both steroids compete for corticosterone binding to the cytosol receptor in vitro and block transport of corticosterone to cell nuclei in vitro in tissue slices. That both steroids are poor blockers of corticosterone cell nuclear translocation in vivo is difficult to reconcile with their antagonistic properties in behavioral effect.

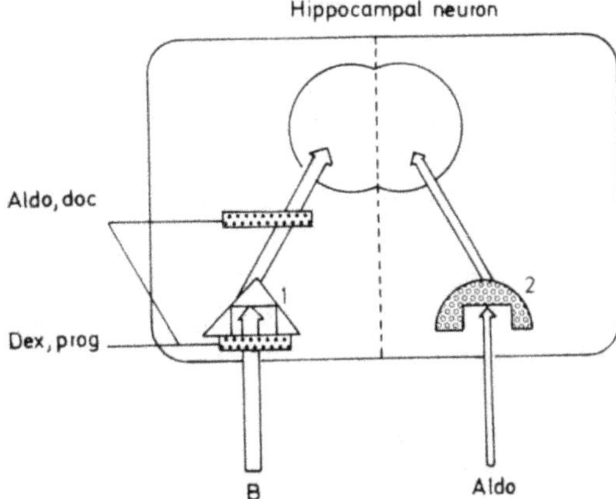

Fig. 1. Schematic representation of gluco- and mineralocorticoid cell interaction. Competition for the binding of corticosterone *(B)* to its receptor site *(1)* occurs in the presence of dexamethasone *(Dex)*, progesterone *(Prog)*, aldosterone *(Aldo)*, and deoxycorticosterone *(Doc)*. Only the latter two steroids (Aldo and Doc) are also able to interfere with the translocation of the steroid-receptor complex from cytoplasma to cell nucleus. Receptor sites for Aldo *(2)* are also present in hippocampal neurons. Whether both gluco- and mineralocorticoid receptor sites coexist in the same neuron or are present in different neurons remains to be established

The deficit in extinction behavior developed 1 h after bilateral removal of the adrenals. Plasma corticosterone levels are then very low; tissue levels are only 6% of that observed 1 h after sham operation, but the cell nuclear compartment still has retained a considerable amount of immunoreactive corticosterone (31% of sham-operated controls). As reported by others (Carroll et al. 1975; Butte et al. 1976; Stevens et al. 1973) plasma and brain tissue levels of corticosterone fluctuate in parallel; the changes in cell nuclear compartment lag behind. The high-affinity "receptor trap" still warrants substantial receptor occupation. Accordingly, receptor occupation has been reported to vary between 50% in the morning and 80% in the evening (McEwen et al. 1974; Turner et al. 1978). Taking binding capacity after 24 h adrenalectomy as 100% (when cytosol corticosterone levels are lower than 15 fmol/mg protein), then at 1 h after adrenalectomy or sham operation 40% or 80% of the sites are still occupied by endogenous steroid (Bohus and de Kloet 1981). It may well be that these binding sites are not only occupied by corticosterone, but also by other antagonist steroids. These steroids may include progesterone, Doc, and aldosterone, but also 19-nor or 18-hydroxyderivatives of corticosterone or Doc and other adrenal steroid intermediates. This had led us to postulate that under physiological conditions a fine-tuned regulation of hippocampal function may exist, which is based on the interplay of the agonist corticosterone and the various antagonists. This proposition does not assign a particular role to the hippocampal corticosterone receptor system. A similar mechanism has been proposed for corticosteroid receptor functioning in the rat kidney and bovine adrenal medulla (Do et al. 1979; Barlow et al. 1979). Antiglucocorticoid action has been described for some of the steroids. Progesterone is an antagonist in various corticosteroid responsive tissues (Samuels and Tomkins 1970; Rousseau et al. 1972; Disorbo et al. 1977) and Doc antagonizes feedback action of corticosterone on pituitary ACTH release (Jones et al. 1976; Duncan and Duncan 1979). Doc also acts as antagonist on fast feedback action (Jones et al. 1974).

Although the experiments with agonist and antagonist steroids on behavior have been the starting point for the postulate of multiple steroid interaction with a single receptor population, it remains to be demonstrated that such a mechanism indeed functions under physiological conditions. In the male rat blood corticosterone level is about 100–1000 times higher than that of aldosterone, progesterone, and Doc, while 18-OH steroids circulate in comparable quantities. Additional experimentation should show how much each of these steroids contributes in vivo to occupation of a steroid receptor system, taking into account such factors as binding to plasma transcortin and receptor heterogeneity (Funder 1977).

The replacement with the behaviorally active dose of 30 μg B/100 g body wt. maintains a receptor occupation close to the level observed for the sham-operated animals at 15 min as well as 60 min as judged from the number of available binding sites detected in vitro with ^{3}H-B (Bohus and de Kloet 1981). This observation suggests that in the given experimental conditions the presence or continuous renewal of B corticosterone receptor complex in the cell nuclear compartment is essential for normal extinction behavior. Corticosterone in a dose of 30 μg/100 g body wt. normalizes the reduced ambulation of these rats up to the level of the sham-operated controls. Receptor occupancy is restored up to the extent of the control animals both at 15 and 60 min after administration, when B is used as the binding ligand. The other steroids result as well in receptor occupancy in the following order of potency B > Doc > Dex > Prog,

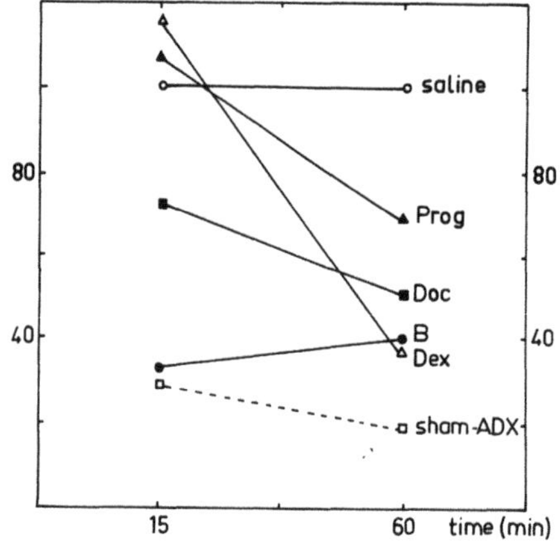

Fig. 2. Receptor occupation in hippocampal cytosol in vitro after in vivo administration of various steroids to adrenalectomized rats. Adrenalectomized rats (10 days) were injected s.c. with steroids in a dose of 30 μg/100 g body wt. 15 or 60 min prior to killing. In vitro receptor occupation was measured using ^{3}H-B as the binding ligand. The binding of ^{3}H-B to hippocampal cytosol of 10-day ADX rats (saline) was taken as 100%. Sham-ADX: sham adrenalectomized rats; saline: 10-day ADX rats

but the temporal pattern is different (Fig. 2). For instance, dexamethasone and progesterone do not occupy B receptor sites 15 min after administration but do so for 62% and 31% respectively after 60 min.

5.2 Neurobiological Studies

Neurobiological mechanisms underlying the corticosterone effects on extinction behavior and exploration should fulfill the same criteria regarding time and dose dependency and steroid and regional specificity as the behavioral effects do. This includes the criterium that neurobiological effects should be manifest within 1 h and respond to low doses of corticosterone. Doc, aldosterone, dexamethasone, and progesterone should display antagonistic effects and, in particular, hippocampal and lateral septal neurons should be responsive. Neurobiological effects observed after gross endocrine and pharmacological manipulations may provide insight into corticosteroid sensitive mechanisms. Only a few aspects will be discussed here. For details the reader is referred to comprehensive reviews by Woodbury et al. 1972; McEwen and Luine 1978; and Rees and Gray 1982.

Any study using adrenalectomy should consider whether the oberved effects are due to removal of the adrenals or due to removal of the adrenocortical feedback on pituitary ACTH release. The 1-h interval after adrenalectomy produces a substantial rise in plasma ACTH level (1076 ± 126 pg/ml vs 499 ± 45 pg/ml sham operated). The 10-day adrenalecomized rats are characterized by a high rate of ACTH production and release leading to very high plasma ACTH levels (1970 ± pg/ml vs 128 ± 17 pg/ml sham operated), when measured immediately after open field exposure. Peptides related to the 31K family of peptides (pro-opiomelanocortin, the precursor of ACTH, MSH, and endorphin) are powerful modulators of neuronal processes underlying the type of behaviors under study (de Kloet and de Wied 1980); one may suspect an interference of these peptides with steroid action in brain. As it has been described previously, such interactions do exist. At this point the first restriction in the mechanism of action can be made. Dexamethasone is a powerful suppressor of ACTH release; yet there is no distinct behavioral effect of the steroid, unless very high amounts are given. Experiments which had been designed to demonstrate behavioral consequence of blocking ACTH release by dexamethasone failed to find behavioral alterations (for detailed discussion see Bohus and de Wied 1980). While this observation minimizes a role of pituitary ACTH, it does not rule out that ACTH-related peptides are effective via the pro-opiomelanocortin neuronal network (Larsson 1978; Watson et al. 1978), which arises from cell bodies in the hypothalamic arcuate nuclei and terminates among others in limbic structures. We have observed in agreement with others, that the level of immunoreactive ACTH in the hippocampus is not affected by hypophysectomy (van Dijk et al. 1981; Krieger et al. 1977). However, after bilateral removal of the adrenals, the tissue is slowly depleted of immunoreactive ACTH. The effect of adrenalectomy even persists after additional removal of the pituitary. The depletion develops slowly and levels are minimal (39.1% of sham-operated controls) after 3 days. Hippocampal ACTH levels are, however, not significantly different at 1 h or 10 days after adrenalectomy. The longer times have been used in general in other studies, which may explain the inability of other workers to find effects of adrenalectomy. In addition, hypothalamic ACTH was not affected. Corticosterone substitution of the adrenalectomized rats did not restore immunoreactive ACTH levels in the hippocampus, indicating that other adrenal factors (cortical and medullary) may be involved in regulation of central ACTH synthesis and/or release.

There is no doubt that corticosteroids have effects on neurotransmitter metabolism either directly or indirectly by altering cell metabolism. Studies on this mode of action have mostly been focused on analyses of rate-limiting enzymes involved in neurotransmitter synthesis or degradation or on neurotransmitter turnover rate. These parameters could be altered via steroid action in the neuronal perikarya, which will result in time delays due to transport of the synthetized product to the cell surface sometimes over long distances via axons to the synaptic endings. On the other hand, the steroid effects could be exerted directly at the pre- or postsynaptic level by affecting membrane properties, high-affinity uptake or binding systems, and biosynthetic or degradative enzyme systems (McEwen et al. 1978).

For attempts to correlate the mechanism of action with the behavioral observations, the time interval at which these effects occur may be of crucial importance. The action of corticosterone may be to maintain or to trigger the necessary genomic processes permitting the behaviors to occur. If indeed genomic processes are involved, the 1-h time interval is not too short. Corticoid effects mediated by genomic events in target tissues such as thymic lymphocytes are reported within 20 min (Makman et al. 1971; Lando and Raynaud 1981). In hippocampal tissue slices corticosterone stimulates ^{3}H-uridine incorporation into cell RNA (Dokas 1979) and ^{3}H-leucine incorporation into specific cytosolic proteins (Etgen et al. 1980). In the latter study progesterone was ineffective and blocked corticosterone effect on protein synthesis. Thus, steroid specificity and dose (10^{-9} M) suggest the involvement of corticosterone receptors in steroid effects on hippocampal protein synthesis.

Catecholamines, indolamines, and GABA have been the most investigated neurotransmitter systems affected by steroids. The number of high-affinity binding sites involved in GABA uptake via synaptic membranes appeared under control of corticosterone (Miller et al. 1978). Although this effect is restricted to hippocampal neurons, its induction is beyond the 1-h time interval. We have studied catecholamine metabolism in the first 2 h after adrenalectomy. We observed that noradrenaline disappearance was increased from the paraventricular nucleus, arcuate nucleus, median eminence, and dorsomedial nucleus of the hypothalamus of rats that had been given α-methyl-p-tyrosin (αMPT) 0.5 h prior to adrenalectomy compared with unoperated, α-MPT pretreated rats. The disappearance of dopamine after synthesis inhibition was increased in the arcuate nucleus of adrenalectomized rats and that of adrenaline in the paraventricular nucleus, arcuate nucleus, and median eminence. In most cases, however, the increases in disappearance following adrenalectomy were not different from sham operation. There were no steroid effects in brain regions containing a large amount of corticosterone receptors (Versteeg et al. 1979). It seems therefore that the observed changes are rather the consequence of the surgical procedure including the ether anesthesia and/or elevated levels of plasma ACTH than of the absence of corticosterone after adrenalectomy.

Serotonin metabolism has been shown to be under the influence of corticosteroids (Kovacs et al. 1977). Chronic adrenalectomy decreases and glucocorticoids increase the formation of serotonin from ^{3}H-tryptophan precursor in vitro and in vivo (Azmitia and McEwen 1969, 1974; Millard et al. 1972; Azmitia et al. 1970). The underlying mechanism may be control of ^{3}H-tryptophan uptake (Neckers and Sze 1975), but also corticoid effects on the uptake and release of serotonin in vitro in synaptosomes have been reported (Vermes et al. 1976), as well directly on tryptophan-hydroxylase (Azmitia and McEwen 1974; Sze et al. 1976).

A dose-dependent dual effect of corticosterone on cerebral 5-HT metabolism was noted (Kovacs et al. 1976). Administration of a low dose of corticosterone (1.0 mg/kg) increases serotonin content and turnover (measured by a MAO inhibitor method), while a high dose (10 mg/kg rat) decreased both parameters. Most studies agree that corticosterone effects on serotonin metabolism have a time lag of 30 min or shorter, that they last not more than a few hours, and that they prevail in hypothalamic and midbrain regions.

In a joint effort with the group of Kovacs and Telegdy in Szeged, Hungary, we have studied the effect of acute adrenalectomy on serotonin metabolism in rats treated 30 min before they were killed with pargyline, a monoamino oxidase (MAO) inhibitor. Since pargyline inhibits degradation of serotonin, this treatment alone results in an accumulation of the indolamine. One hour after adrenalectomy the pargyline-induced accumulation was reduced in the dorsal hippocampus and raphe area, while the hypothalamus and septal area were not significantly affected. Replacement with a behaviorally effective dose of corticosterone normalized the reduced serotonin accumulation up to the level of the sham-operated controls in the hippocampus and raphe area. Dexamethasone administration to adrenalectomized rats was ineffective or even reduced serotonin accumulation in septum and hippocampus (de Kloet et al., unpublished). These effects on one aspect of serotonin metabolism therefore show a striking parallel with the characteristics of the corticosterone receptors in the hippocampus and the corticosterone effects on behavior. Whether modulation of serotonin metabolism is required for the expression of corticosterone effects on extinction behavior or whether altered serotonin metabolism is correlated with the behavioral adaptation is presently under investigation.

6 Modulation of Corticosterone Receptor Capacity

Receptor binding and cell nuclear retention mechanism in the hippocampus, but also in other brain regions, vary through lifespan. The ontogeny is characterized by the lowest number of receptor sites on postnatal day 1. The cytosol binding capacity increases fourfold and gradually reaches adult levels around 4 weeks of age (Olpe and McEwen 1976; Clayton et al. 1977). Development of cell nuclear retention follows a similar time course (Turner 1978) and is in parallel with the genesis of the dentate gyrus neurons (Altman and Bayer 1975). Hypothalamic cytosol binding and cell nuclear uptake of corticosterone are changed to a much lesser extent.

Strikingly different is the ontogeny of glucocorticoid receptors in the anterior pituitary. While cytosol binding for corticosterone equals that of dexamethasone the first postnatal week, only the capacity for corticosterone shows a further increase, reaching maximal levels after 4 weeks (Olpe and McEwen 1976). In contrast, cell nuclear uptake of a tracer dose of tritiated B is highest the first postnatal week and then gradually decreases (Turner 1978). It was suspected that the TL binding system in the pituitary (Sect. 3.4) developed only after the first postnatal week and indeed the concentration of plasma transcortin or the TL binding system in the pituitary was very low the 1st week and even undetectable between day 6 and day 10 (Sakly and Koch 1981). Accordingly, this finding further supports a physiological role of the TL binding system in that it blunts changes in plasma B level for direct interaction with glucocorticoid receptors in the pituitary (de Kloet et al. 1977a, b; Koch et al. 1977).

It is of interest to note that the pituitary-adrenal system displays changes in activity, which can be related to the animal's age. Although at every time in the newborn rat a stress response can be provoked, provided the stimulus is strong enough, the responsiveness is not fully developed before the second postnatal week (Schapiro 1968; Hiroshige and Sato 1970; Levine 1970; Gray 1971). Circadian rhythmicity gradually appears as the animal matures and is found developed at approximately 3 weeks (Ader 1968; Hiroshige and Sato 1970). It has also been found that the development of pituitary-adrenal function is accelerated by handling or by exposing the neonatal rats to a moderate electric shock. Moreover, these procedures have been found to have consequences for behavior and adrenal function in adulthood (Ader 1968; Levine et al. 1967). Senescent rats, alternatively, have elevated plasma corticosterone levels and heavier adrenals (Tang and Phillips 1978). This may be related to a progressive astrogliosis, which was observed in aging rats, and presumably leads to disinhibitory influences on hippocampus function (Landfield et al. 1978).

There is a striking parallelism between these features of pituitary-adrenal function and postnatal neurogenesis in hippocampus and appearance of receptor sites in hippocampus, and the progressive hippocampal astrogliosis (Landfield et al. 1978) and disappearance of receptor sites during aging. If indeed in the hippocampus as in other steroid-cell systems the capacity of the receptors is the limiting factor in the ability of the target cell to respond, then it is very important to know how the number of receptor sites is regulated. At present we are able to distinguish three categories of corticosterone receptor capacity control in the rat hippocampus (Veldhuis and de Kloet 1981; de Kloet et al. 1980). These are autoregulation by corticosterone, influences by neurotropic factors, and the responses observed after lesions of hippocampal inputs.

The first, autoregulation by corticosterone, represents a mechanism where excess corticosterone regulates down the capacity of its own receptor (McEwen et al. 1974; Valeri et al. 1978; de Kloet and Veldhuis 1980). Autoregulation by corticosterone becomes manifest after removal of the adrenals, after chronic corticosterone treatment, or in chronically stressed animals. After adrenalectomy there is a rapid rise caused by depletion of endogenous steroid from previously occupied receptor sites followed by a more gradual increase in receptor number (McEwen et al. 1974; de Kloet and Veldhuis 1980). It is conceivable that the decreased receptor capacity observed in the senescent mice is related to increased corticosterone levels (Roth 1976).

In the second category are neurotropic substances, such as the neuropeptides related to ACTH and the neurohypophyseal hormones. The implication of these neuropeptides in receptor capacity control has become apparent after classical endocrine approaches: removal of an endocrine gland (pituitary, adrenal) or using a rat strain with a genetically determined lack in vasopressin (Brattleboro rats are homozygous for hypothalamic diabetes insipidus).

Removal of the pituitary resulted in a gradual increase in receptor capacity of hippocampus cytosol, which exceeds that after bilateral removal of the adrenals after 2 weeks with 30%. Since hypophysectomized animals still have adrenals present, the higher capacity observed in these animals was the first indication that more factors than solely the adrenal steroids are involved in receptor capacity control (Table 8). Substitution every other day with ACTH 4–10, a behaviorally active ACTH sequence (de Wied 1969) decreased receptor capacity to the level of the sham-operated control animals (Veldhuis and de Kloet 1982a). ACTH 4–10 is virtually devoid of corticotropic activity and thus represents a direct effect of the peptide on the brain.

Table 8. Effect of adrenalectomy (ADX) and hypophysectomy (hypox) on in vitro ^{3}H-corticosterone binding[a] to hippocampal cytosol.

Treatment	B max	Kd.
	(fmol/mg prot.) Mean ± SEM	(nM) Mean ± SEM
Sham hypox 24 h ADX	304 ± 13.3 [b]	2.6 ± 0.4
Sham hypox 14 day ADX	383 ± 8.1 [c]	2.0 ± 0.2
14 day hypox 24 h ADX	497 ± 12.0 [c]	2.3 ± 0.2

[a] ^{3}H-corticosterone binding to hippocampal cytosol was measured using different concentrations (0.5–30 μM) of the steroid. Scatchard plot analysis was used to calculate the maximal binding capacity (B max) and dissociation constant (Kd).

[b] n = 7.

[c] $P < 0.001$ vs sham hypox, 24 h ADX.

The Brattleboro rat, homozygous for diabetes insipidus, has an approximately 30% lower receptor capacity for corticosterone than nondiabetes homozygous animals of the same strain (de Kloet and Veldhuis 1980). These animals have a hereditary lack in the synthesis of vasopressin, which is clearly manifest in the inability to retain water (Valtin and Schroeder 1964). Deficits in the acquisition and maintenance of new behavioral patterns have also been found (de Wied et al. 1975; Bohus et al. 1975). Administration of arginine-vasopressin every day for 1 week increased receptor capacity up to the level of the nondiabetes control animals (Veldhuis and de Kloet 1982b).

Desglycinamide-arginine-vasopressin, a peptide lacking the antidiuretic activity of vasopressin was also effective. Both peptides improve the impaired avoidance behavior of diabetes rats (de Wied et al. 1975).

In the third category are the responses observed after lesioning neural connections of the hippocampus. It became first apparent in a study where the effect of discrete septal lesions on corticosterone binding in the hippocampus was investigated, while the lesions were functionally characterized by a number of parameters such as hippocampal theta activity, behavioral response, sensitivity to electric footshock, avoidance behavior, and pituitary-adrenal activity (Nyakas et al. 1979). Only lesions in the dorsal septal area, a region known to be rich in corticosterone receptor sites (Gerlach and McEwen 1972), altered the receptor capacity in the hippocampus. The lesion increased the number of cytosol receptor sites in the hippocampus. Another outcome of the study was that the acquisition of conditioned avoidance response was severely impaired, suggesting that full integration of the septal-hippocampal complex through efferents from the hippocampus to the dorsal septum is necessary for optimal expression of the behavior. Other lesions, for instance, in the parafascicular nucleus, impair acquisition of an avoidance response as well, but do not alter hippocampal B receptor capacity. We have postulated the existence of a compensatory increase in remaining corticosterone receptor containing neurons after removal of part of the system (Nyakas et al. 1979). Indeed, when the dorsal hippocampus is removed unilaterally, the receptor capacity increases considerably in the remaining contralateral lobe (Nyakas et al. 1981).

At this point it is difficult to envisage what the functional meaning of such experimentally induced changes in receptor capacity is. Dose-response studies on hippocampal dependent behaviors such as extinction or exploration behavior, or electrophysiological parameters may provide an answer to this question. The change in receptor capacity observed after the lesion studies may even be related to axonal growth or glial proliferation since extensive degeneration and regeneration of nerve tissue is known to occur (McWilliams and Lynch 1980).

It is of interest to note that there is a large variability in corticosterone receptor binding capacity in naive rats of the same age and which are maintained under apparently homogenous conditions. Individual binding values of such rats ranged between 200 and 550 fmol/mg protein in hippocampus cytosol. Classification of these animals in groups displaying good or poor passive avoidance behavior or acquisition of a conditioned escape response is paralleled by a significant difference in binding capacity (Angelucci et al. 1980). It is possible that these individual differences are not only genetically determined, but also the consequence of factors impinging on the receptor system for corticosterone in the hippocampus during lifespan. Such factors are possibly of environmental origin and may act via hormonal and neuropeptide factors. Hippocampal formation in its early stages of postnatal development might be particularly susceptible to these influence and the corticosterone receptor system might be one of the systems reflecting hippocampus function in later life.

7 General Discussion

The pituitary-adrenal system plays an essential role in the physiological control of adaptive behavior through its secretion of adrenocortical hormones. These steroid hormones are a class of potent modulators of brain function. The steroid action appears complex and appears to include behavioral and neuroendocrinological components. On the basis of experimental design (endocrine status, dose of steroids, type of behavioral assay) and the investigated ligand, it is now possible to recognize distinct steroid actions on the brain and to evaluate the role of the corticosterone receptor system in the modulation of hippocampal function.

The first dissociation is that between neuroendocrine regulation and behavior. Potent glucocorticoids are in general potent suppressors of stress-induced ACTH release. Alternatively, we have found that in this respect less potent, but naturally occurring glucocorticoid of the rat, corticosterone, is among the steroids the most specific and potent modulator of neuronal processes underlying adaptive behavior.

The second dissociation is closely related to the first. The sites, where behavioral and neuroendocrine regulations are primarily affected, are distinctly different. Potent synthetic and natural glucocorticoids act on the hypothalamo-hypophyseal complex to regulate pituitary ACTH release. Modulation of behavior by corticosterone takes place via extrahypothalamic structures, particularly in the hippocampus. The ligand specificity of the receptor sites is in agreement with the dissociation in steroid actions.

The third dissociation becomes apparent from the selected experimental design to study steroid action and the subsequent status of the experimental subjects. What may be indicated as a pharmacological type approach has been shown to provoke profound changes in behavior. Irrespective of the glucocorticoid, the general effect was to facilitate extinction and suppress retention behavior in the intact rat. It is not yet clear what the mechanism of action is in this rather nonspecific steroid effect on behavior and whether steroid receptor sites of the type described in this chapter are involved. Using a *classical endocrine* approach we have been able to specify steroid actions on behavior. This approach leaves corticosterone as the only agonist for physiological control of extinction behavior; the steroid facilitates such behavior. All other pregnene steroids which were shown active agonists in the pharmacological studies appear inactive and even block corticosterone action under experimental conditions designed to evaluate its physiological significance and its relation to the hippocampus.

A common background of these three dissociations may well be a differential interaction of corticosteroids with the two pro-opio-melanocortin system of brain and pituitary. In both tissues the 31K precursor molecule is found, but present evidence suggests that the two systems are separately regulated and have different end products formed by enzymatic cleavage (Eipper and Mains 1978; Watson and Akil 1980). The release of ACTH and β-endorphin is concomitant and modulated by the same stimuli, e.g., stress, corticotropin-releasing factor, and vasopressin (Guillemin et al. 1977; Allen et al. 1978; Vale et al. 1978; Przewlocki et al. 1979). In addition, like ACTH, synthesis and release of β-endorphin in the anterior pituitary is increased by adrenalectomy and suppressed by glucocorticoids, such as dexamethasone and corticosterone (Guillemin et al. 1977; Vale et al. 1978; Rossier et al. 1979; Przewlocki et al. 1979; Simantov 1979; Allen et al. 1978). The release of the end product peptides of the brain pro-opio-melanocortin system has been minimally studied. It may well be that corticosteroids acting either at the cell bodies or terminal levels affect synthesis, release, and degradation of β-endorphin, ACTH, and/or α-MSH in the brain. Interactions between corticosteroids and endorphins, which are highly potent modulators of behavioral adaptation (de Wied 1979; Bohus 1980; de Kloet and de Wied 1980) have not yet been explored. Interactions between corticosteroids and ACTH-related peptides on the behavior of intact and adrenalectomized rats have been observed.

Involvement of an endocrine feedback, i.e., suppression of ACTH release, has been excluded as a major mechanism in the behavioral effects of adrenal steroids in the intact rat. Structural requirements for certain behavioral effects in the intact rat and delayed corticosteroid feedback on ACTH release at hypothalamic level (Bohus and Strashimirov 1970; Jones et al. 1977) are, however, rather similar. Whether this similarity is an indication of an interaction between corticosteroids and brain pro-opiomelanocortin fragments remains an open question.

The three dissociations have enabled us to relate corticosterone actions to the limbic system. The majority of the receptor sites with stringent specificity for corticosterone are localized in the hippocampus. It is therefore reasonable to designate (forward) the hippocampus as the major target of corticosterone-specific behavioral changes. In line with this reasoning one could view hippocampal functioning as determined by corticosterone-receptor interaction. To compare, however, the action of increasing doses of corticosterone on hippocampal function with one approaching the effect of hippocampal ablation, as suggested by McEwen (1980), in our opinion may lead to oversimplification of a modulatory role of corticosterone in adaptive behavior. The number of behavioral changes that follow hippocampectomy is much more substantial than those induced by adrenalectomy (see, e.g., Micco and McEwen 1980). It is therefore more likely that certain hippocampal functions are more related to a corticosterone-receptor interaction than others. In order to decide which hippocampal functions would fall into the above category, a reinvestigation of the behavioral effects of corticosterone in relation to a) steroid specificity and b) the recent knowledge of hippocampal functioning seems to be imperative.

The most recent behavioral studies were guided first of all by what was discovered on the interaction of adrenocortical hormones with receptor systems in brain and pituitary. The findings on agonist/antagonist relationships in steroid regulation of behavior are tentatively explained in the receptor model presented in Fig. 1. There are specific neuronal corticosterone receptor sites (apart from those for mineralocorticoids) and corticosterone is a powerful promotor in cell nuclear translocation of the receptor complex. A number of other steroids interfere with corticosterone binding at various levels and block the cell nuclear uptake process. It is conceivable that under physiological conditions a finely tuned regulation exists, which is based on the relative available amounts of each of the steroids. A subsequent proportional receptor occupation by corticosterone leads to a signal triggering genomic events.

Although at this point a great step forward has been made in understanding the molecular mechanism underlying endocrine control of brain functions, the sequence of biochemical events taking place after steroid-receptor interaction is still poorly understood. What may turn out, however, as the principal mediator of the modulatory action of corticosterone on neural processes is the serotonergic system. Our most recent experiments have revealed a close parallel of an index for serotonin turnover with the agonistic action of the steroid on extinction behavior. The development of a behavioral deficit shortly after adrenalectomy is correlated with a reduction in pargyline-induced accumulation of the indolamine in the dorsal hippocampus and the raphe area. Both parameters are restored by corticosterone, but not by dexamethasone. The latter steroid even blocks the normalizing effect of corticosterone on serotonin turnover. What the actual hierarchy is in the correlates of behavior with serotonin and corticosterone receptor interaction as the respective neural and endocrine components remains to be established. It could well be that any disturbance in either one of these parameters triggers responses from the other systems in order to maintain a certain "state" in hippocampal functioning. In this sense corticosterone action under steady state conditions could be interpreted biochemically as action to maintain a given rate of cell metabolism.

Modulation of hippocampal functioning then on the one hand depends on the pattern of steroids and the level of corticosterone the tissue is exposed. On the other hand, the tissue receptor system for corticosterone may modulate hippocampal functioning, and different levels of activity may be explained by different receptor capacities. Using an endocrine approach, it was shown that the pituitary-adrenal system via corticosterone, as well as neuropeptides related to the opiocortins and to neurohypophyseal hormones, regulates hippocampal corticosterone receptor capacity.

Lesions leading to rapid degeneration and regeneration in the hippocampus, such as those occurring after dorsal septal lesions or contralateral hippocampectomy, also lead to pronounced changes in hippocampal corticosterone receptor capacity. These changes may indicate an alteration in the capacity of the neuronal receptor system or may be a function of the enhanced glial cell proliferation. Therefore, one of the primary goals would be to resolve the cellular compartimentalization in glucocorticoid receptor sites in order to assign receptor capacity changes to neurons or glial cells after various experimentally induced manipulations.

What comes as a sobering thought in evaluation of such induced receptor capacity changes is that the most pronounced alterations occur during lifespan. The increase in capacity during early life and a subsequent decrease in the senescent rat may just reflect the resultant of all environmental, neuronal, and endocrine factors impinging on hippocampus function and lead to that receptor capacity for corticosterone best fitted to cope with the needs of the *milieu exterieur et interieur.*

Acknowledgments. Parts of the studies were supported by the Stichting Pharmacologisch Studiefonds and the Dutch-Soviet Cultural Exchange Program of the Ministry of Education and Sciences. H.D.V. was supported by the Foundation for Medical Research (FUNGO), which is subsidized by the Netherlands Organization for the Advancement of Pure Research (ZWO).

The authors are indepted for the valuable technical assistance of Miss Anja ter Avest.

The steroids and peptides were donated by Organon B.V., Oss, The Netherlands.

References

Ader R (1968) Effects of early experience on emotional and physiological reactivity in the rat. J Comp Physiol Psychol 66:264–268

Ader R, Weijnen JAWM, Moleman P (1972) Retention of a passive avoidance response as a function of the intensity and duration of electric shock. Psychon Sci 26:125–128

Agawal MK (1978) Physical characterization of cytoplasmic gluco- and mineralocorticoid receptors. FEBS Lett 85:1–8

Allen RG, Herbert E, Hinman M, Shibuya H, Pert CB (1978) Coordinate control of corticotropin, β-lipotropin and β-endorphin release in mouse pituitary cell-cultures. Proc Natl Acad Sci USA 75:4972–4976

Altman G, Bayer S (1975) Postnatal development of the hippocampal dentate gyrus under experimental conditions. In: Isaacson RL, Pribram KH (eds) The hippocampus, vol I. Plenum Press, New York, pp 95–122

Anderson NS, Fanestil DD (1976) Corticoid receptors in rat brain: evidence for an aldosterone receptor. Endocrinology 98:676–684

Angelucci L, Valeri P, Palmery M, Patacchioli FR, Catalani A (1980) Brain glucocorticoid receptor: correlation of in vivo uptake of corticosterone with behavioral, endocrine and neuropharmacological events. In: Pepeu G, Kuhar MJ, Enna SJ (eds) Receptors for neurotransmitters and peptide hormones. Raven Press, New York, pp 391–406

Azmitia EC Jr, McEwen BS (1969) Corticosterone regulation of tryptophan hydroxylase in midbrain of the rat. Science 166:1274–1276

Azmitia EC Jr, McEwen BS (1974) Adrenocortical influence on rat tryptophan hydroxylase activity. Brain Res 78:291–302

Azmitia EC Jr, Algeri S, Costa E (1970) Turnover rate of in vivo conversion of tryptophan into serotonin in brain areas of adrenalectomized rats. Science 196:201–203

Ballard PL, Ballard RA (1974) Cytoplasmic receptor for glucocorticoids in lung of human fetus and neonate. J Clin Invest 53:477–486

Barlow JW, Kraft N, Stockigt JR, Funder JW (1979) Predominant high affinity binding of [^{3}H]-dexamethasone in bovine tissues is not to classical glucocorticoid receptors. Endocrinology 105:827–834

Baxter JD, Rousseau GG (1979) Glucocorticoid hormone action. Springer, Berlin Heidelberg New York

Beatty PA, Beatty WW, Bowman RE, Gilchrist JC (1970) The effects of ACTH, adrenalectomy and dexamethasone on the acquisition of an avoidance response in rats. Physiol Behav 5:939–944

Berthold K, Arimura A, Schally AV (1970) In vivo studies on the mechanism of action of 6-dehydro-16-methylene-hydrocortisone (STC 407) on the hypothalamo-pituitary-adrenal axis in the rat. Neuroendocrinology 6:301–310

Bohus B (1968) Pituitary ACTH release and avoidance behaviour of rats with cortisol implants in mesencephalic reticular formation and median eminence. Neuroendocrinology 3:355–365

Bohus B (1970a) Central nervous structures and the effect of ACTH and corticosteroids on avoidance behaviour: a study with intracerebral implantation of corticosteroids in the rat. Prog Brain Res 32:171–184

Bohus B (1970b) The medial thalamus and the opposite effect of corticosteroids and adrenocorticotrophic hormone on avoidance extinction in the rat. Acta Physiol Acad Sci Hung 38:217–223

Bohus B (1971) Adrenocortical hormones and central nervous function: The site and mode of their behavioural action in the rat. In: James VHT, Martini L (eds) Hormonal steroids. Proc Third Int Congr Hormonal Steroids, Series no 219. Excerpta Medica, Amsterdam, pp 752–758

Bohus B (1973) Pituitary-adrenal influences on avoidance and approach behavior of the rat. Prog Brain Res 39:407–420

Bohus B (1974) Pituitary-adrenal hormones and the forced extinction of a passive avoidance response in the rat. Brain Res 66:366–367

Bohus B (1975) The hippocampus and the pituitary-adrenal system hormones. In: Isaacson RL, Pribram KH (eds) The hippocampus, vol 1. Plenum Press, New York, pp 323–353

Bohus B (1979) Effects of ACTH-like neuropeptides on animal behavior and man. Pharmacology 18:113–122

Bohus B (1980) Endorphins and behavioral adaptation. Adv Biol Psychiatr 5:7–19

Bohus B, De Kloet ER (1977) Behavioural effects of corticosterone related to putative glucocorticoid receptor properties in the rat brain. J Endocrinol 72:64P–65P

Bohus B, De Kloet ER (1981) Adrenal steroids and extinction behavior: Antagonism by progesterone deoxycorticosterone, and dexamethasone of a specific effect of corticosterone. Life Sci 28:433–440

Bohus B, De Wied D (1966) Inhibitory and facilitatory effect of two related peptides on extinction of avoidance behavior. Science 153:318–320

Bohus B, De Wied D (1967) Failure of α-MSH to delay extinction of conditioned avoidance behavior in rats with lesions in the parafascicular nuclei of the thalamus. Physiol Behav 2:221–223

Bohus B, De Wied D (1980) Pituitary-adrenal system hormones and adaptive behaviour. In: Chester-Jones I, Henderson IW (eds) General, comparative and clinical endocrinology of the adrenal cortex, vol 3. Academic Press, London, pp 256–347

Bohus B, Endröczi E (1965) The influence of pituitary-adrenocortical function on the avoiding conditioned reflex activity in rats. Acta Physiol Acad Sci Hung 26:183–189

Bohus B, Lissak K (1968) Adrenocortical hormones and avoidance behaviour of rats. Int J Neuropharmacol 7:301–306

Bohus B, Strashimirov D (1970) Localization and specificity of corticosteroid "feedback receptors" at the hypothalamo-hypophyseal level; comparative effect of various steroids implanted in the median eminence or the anterior pituitary of the rat. Neuroendocrinology 6:197–209

Bohus B, Nyakas Cs, Endröczi E (1968) Effects of adrenocorticotropic hormone on avoidance behaviour of intact and adrenalectomized rats. Int J Neuropharmacol 7:307–314

Bohus B, Grubits J, Kovacs G, Lissak K (1970) Effect of corticosteroids on passive avoidance behaviour of the rat. Acta Physiol Acad Sci Hung 38:381–391

Bohus B, Van Wimersma Greidanus TjB, De Wied D (1975) Behavioral and endocrine responses of rats with hereditary hypothalamic diabetes insipidus (Brattleboro strain). Physiol Behav 14:609–615

Bookin HB, Pfeiffer WD (1978) Adrenalectomy attenuates electroconvulsive shock-induced retrograd amnesia in rat. Behav Biol 24:527–532

Buckingham JC, Hodges JR (1974) Interrelationships of pituitary plasma corticosterone in adrenalectomized and stressed, adrenalectomized rats. J Endocrinol 63:213–222

Butte JC, Kakihana R, Noble EP (1976) Circadian rhythm of corticosterone levels in rat brain. J Endocrinol 68:235–239

Carroll BJ, Heath B, Jarrett DB (1975) Corticosteroids in brain tissue. Endocrinology 97:290–300

Clayton CJ, Grosser BI, Stevens W (1977) The ontogeny of corticosterone and dexamethasone receptors in rat brain. Brain Res 134:445–453

Conner RL, Levine S (1969) The effects of adrenal hormones on the acquisition of signaled avoidance behavior. Horm Behav 1:73–83

Dallman MF, Yates FE (1969) Dynamic asymetrics in the corticosteroid feedback path and distribution-metabolism-binding elements of the adrenocortical system. Ann NY Acad Sci 156:696–721

Dallman MF, Demanincor D, Shinsako J (1974) Diminishing corticotrope capacity to release ACTH during sustained stimulation: the twenty-four hours after bilateral adrenalectomy in the rat. Endocrinology 95:65–73

De Kloet ER, Burbach JPH (1978) Selective purification of a single population of glucocorticoid receptors from rat brain. J Neurochem 30:1505–1507

De Kloet ER, De Wied D (1980) The brain as target tissue for hormones of pituitary origin: Behavioural and biochemical studies. Front Neuroendocrinol 6:157–201

De Kloet ER, McEwen BS (1976) Differences between cytosol receptor complexes with corticosterone and dexamethasone in hippocampal tissue from rat brain. Biochim Biophys Acta (Amst) 421:124–132

De Kloet ER, Veldhuis HD (1980) The hippocampal corticosterone receptor system of the homozygous diabetes insipidus (Brattleboro) rat. Neurosci Lett 16:187–192

De Kloet ER, Van der Vies J, De Wied D (1974) The site of suppressive action of dexamethasone on pituitary-adrenal activity. Endocrinology 94:61–73

De Kloet ER, Wallach G, McEwen BS (1975) Differences in corticosterone and dexamethasone binding to rat brain and pituitary. Endocrinology 96:598–609

De Kloet ER, Burbach P, Mulder GH (1977a) Localization and role of transcortin-like molecules in the anterior pituitary. Mol Cell Endocrinol 7:261–273

De Kloet ER, Dam CW, Bohus B (1977b) Multiplicity of binding systems specific for glucocorticoids in rat brain and pituitary. In: Agarwal MK (ed) Multiple molecular forms of steroid hormone receptors. Elsevier/North Holland, Amsterdam, pp 65–79

De Kloet ER, Veldhuis HD, Bohus B (1980) Significance of neuropeptides in the control of corticosterone receptor activity in rat brain. In: Pepeu G, Kuhar MJ, Enna SJ (eds) Receptors for neurotransmitters and peptide hormones. Raven Press, New York, pp 373–382

De Vellis J, McEwen BS, Cole R, Inglish D (1974) Relations between glucocorticoid nuclear binding, cytosol receptor activity and enzyme induction in a rat glial cell line. J Steroid Biochem 5:392–393

De Wied D (1967) Opposite effects of ACTH and glucocorticosteroids on extinction of conditioned avoidance behavior. Exerpta Medica Int Congr Series 132:945–951

De Wied D (1969) Effect of peptide hormones on behavior. Front Neuroendocrinol pp 97–140

De Wied D (1979) Schizophrenia as an inborn error in the degradation of β-endorphin – a hypothesis. Trends in Neurosci 2:79–82

De Wied D, Bohus B, Van Wimersma Greidanus TjB (1975) Memory deficit in rats with hereditary diabetes insipidus. Brain Res 85:152–156

Di Sorbo D, Rosen F, McPhartland RP, Milholland RJ (1977) Glucocorticoid activity of various progesterone analogs: correlations between specific binding in thymus and liver and biological activity. Ann NY Acad Sci 286:355–368

Do YS, Loose DS, Feldman D (1979) Heterogeneity of glucocorticoid binders: A unique and a classical dexamethasone-binding site in bovine tissue. Endocrinology 105:1055–1063

Dokas LA (1979) Corticosterone and RNA metabolism in the rat hippocampus. Soc Neurosc (Abstr) 5:443

Douglas RJ (1967) The hippocampus and behavior. Psychol Bull 67:416–422

Duncan MR, Duncan GR (1979) An in vivo study of the action of antiglucocorticoids on thymus weight ratio, antibody titre and the adrenal-pituitary-hypothalamus axis. J Steroid Biochem 10:245–259

Dunn AJ (1980) Neurochemistry of learning and memory: an evaluation of recent data. Annu Rev Psychol 31:343–390

Eipper BA, Mains RE (1978) Existence of a common precursor to ACTH and endorphin in the anterior and intermediate lobes of the rat pituitary. J Supramol Struct 8:247–262

Endröczi E (1972) Limbic system, learning and pituitary-adrenal function. Akademiai Kiado, Budapest

Ermisch A, Rühl H-J (1978) Autoradiographic demonstration of aldosterone-concentrating neuron populations in rat brain. Brain Res 147:154–158

Etgen AM, Martin M, Gilbert R, Lynch G (1980) Characterization of corticosterone-induced protein synthesis in hippocampal slices. J Neurochem 35:598–602

Feldman D, Funder J, Loose D (1978) Is the glucocorticoid receptor identical in various target organs? J Steroid Biochem 9:141–145

Flexner JB, Flexner LB (1970) Adrenalectomy and the suppression of memory by puromycin. Proc Natl Acad Sci USA 66:48–52

Flood JF, Vidal D, Bennet EL, Orme AE, Vasquez S, Jarvik ME (1978) Memory facilitating and anti-amnesic effects of corticosteroids. Pharmacol Biochem Behav 8: 81–87

Fuller JL, Chambers RM, Fuller RP (1956) Effects of cortisone and of adrenalectomy on activity and emotional behavior of mice. Psychosom Med 29:323–328

Funder JW (1977) Multiplicity of steroid-hormone receptors: a physiological role? In: Agarwal MK (ed) Multiple molecular forms of steroid hormone receptors. Elsevier/North Holland Biomedical Press, Amsterdam, pp 263–279

Funder JW, Feldman D, Edelman IS (1973a) The roles of plasma binding and receptor specificity in the mineralocorticoid action of aldosterone. Endocrinology 92: 994–1004

Funder JW, Feldman D, Edelman IS (1973b) Glucocorticoid receptors in rat kidney: the binding of tritiated-dexamethasone. Endocrinology 92:1005–1013

Gerlach JL, McEwen BS (1972) Rat brain binds adrenal steroid hormone: radioautography of hippocampus with corticosterone. Science 175:1133–1136

Gerlach JL, McEwen BS, Pfaff DW, Moskovitz S, Ferin M, Carmel PW, Zimmerman EA (1976) Cells in regions of rhesus monkey brain and pituitary retain radioactive estradiol, corticosterone and cortisol differentially. Brain Res 103:603–612

Gray P (1971) Pituitary-adrenal response to stress in the neonatal rat. Endocrinology 89:1126–1128

Gray P (1976) Effect of prestimulation on avoidance responding in rats, and hormonal dependence of the effect. J Comp Physiol Psychol 90:1–17

Grosser BI, Stevens W, Reed PJ (1973) Properties of corticosteroid-binding macromolecules from rat brain cytosol. Brain Res 57:387–396

Guillemin R, Vargo T, Rossier J, Minick S, Ling N, Rivier C, Vale W, Bloom F (1977) β-endorphin and adrenocorticotropin are secreted concomitantly by the pituitary gland. Science 197:1367–1369

Gyermek L, Genther G, Fleming N (1967) Some effects of progesterone and related steroids on the central nervous system. Int J Neuropharmacol 6:191–198

Harrison RW, Fairfield S, Orth DN (1977) The effect of membrane alteration on glucocorticoid uptake by the AtT-20 target cell. Biochim Biophys Acta 466:357–364

Hennessy JW, Cohen ME, Rosen AJ (1973) Adrenocortical influences upon the extinction of an appetitive runway response. Physiol Behav 11:767–770

Hennessy JW, Smotherman WP, Levine S (1976) Conditioned taste aversion and the pituitary-adrenal system. Behav Biol 16:413–424

Hennessy JW, Smotherman WP, Levine S (1980) Investigations into the nature of the dexamethasone and ACTH effects upon learned taste aversion. Physiol Behav 24: 645–649

Heybach JP, Coover GD, Lints CE (1978) Behavioral effects of neurotoxic lesions of the ascending monoamine pathways in the rat brain. J Comp Physiol Psychol 92: 58–70

Hiroshige T, Sato T (1970) Circadian rhythm and stress-induced changes in hypothalamic content of corticotrophin releasing activity during postnatal development in the rat. Endocrinology 86:1184–1186

Joffe JM, Mulick JA, Rawson RA (1972) Effects of adrenalectomy on open-field behavior in rats. Horm Behav 3:87–96

Jones MT, Tiptaft EM, Brush FR, Fergusson DAN, Neame RLB (1974) Evidence for dual corticosteroid-receptor mechanisms in the feedback control of adrenocortico-trophin secretion. J Endocrinol 60:223–233

Jones MT, Hillhouse E, Burden J (1976) Secretion of corticotropin-releasing hormone in vitro. Front Neuroendocrinol 4:195–226

Jones MT, Hillhouse EW, Burden JL (1977) Structure-activity relationships of cortico-steroid feedback at the hypothalamic level. J Endocrinol 74:415–424

Kimble DP (1968) Hippocampus and internal inhibition. Psychol Bull 70:285–295

Koch B, Lutz B, Briaud B, Mialhe C (1976) Heterogeneity of pituitary glucocorticoid binding. Evidence for a transcortin-like compound. Biochim Biophys Acta (Amst) 444:497–507

Koch B, Lutz-Bucher B, Briaud B, Mialhe C (1977) Glucocorticoid binding to plasma membranes of the adenohypophysis. J Endocrinol 73:399–400

Koch B, Lutz-Bucher B, Briaud B, Mialhe C (1978) Inverse effects of corticosterone and thyroxine on glucocorticoid binding sites in the anterior pituitary gland. Acta Endocrinol 88:29–37

Kovacs GL, Telegdy G (1978) Indolamines and behavior: The possible role of seroton-ergic mechanisms in the pituitary-adrenocortical hormone induced behavioral action. In: Lissak K (ed) Results in neuroendocrinology, neurochemistry and sleep research, vol 7. Akademiai Kiado, Budapest, pp 31–97

Kovacs GL, Telegdy G, Lissak K (1976) 5-Hydroxytryptamine and the mediation of pituitary-adrenocortical hormones in the extinction of active avoidance behaviour. Psychoneuroendocrinology 1:219–230

Kovacs GL, Telegdy G, Lissak K (1977) Dose-dependent action of corticosteroids on brain serotonin content and passive avoidance behavior. Horm Behav 8:155–165

Kraulis I, Foldes G, Traikov H, Dubrovsky B, Birmingham MK (1975) Distribution, metabolism and biological activity of deoxycorticosterone in the central nervous system. Brain Res 88:1–14

Krieger DT, Liotta A, Suda T, Palkovits M, Brownstein MJ (1977) Presence of immuno-assayable β-lipotropin in bovine brain and spinal cord: lack of concordance with ACTH concentration. Biochim Biophys Res Comm 16:930–936

Landfield PW, Waymire JC, Lynch G (1978) Hippocampal aging and adrenocorticoids: Quantitative correlation. Science 202:1098–1101

Larsson L-I (1978) Distribution of ACTH-like immunoreactivity in rat brain and gastro-intestinal tract. Histochemistry 55:225–233

Levine S (1968) Hormones and conditioning. In: Arnold WJ (ed) Nebraska symposium on motivation. University of Nebraska Press, Lincoln, Nebraska, pp 85–101

Levine S (1970) The pituitary-adrenal system and the developing brain. Progr Brain Res 32:79–85

Levine S, Haltmeyer GC, Kara GC, Denenberg VH (1967) Physiological and behavioral effects of infantile stimulation. Physiol Behav 2:55–59

MacLusky NJ, Turner BB, McEwen BS (1977) Corticosteroid binding in rat brain and pituitary cytosols: resolution of multiple binding components by polyacrylamide gel based isoelectric focusing. Brain Res 130:564–571

Makman MH, Dworkin B, White A (1971) Evidence for induction by cortisol in vitro of a protein inhibitor of transport and phosphorylation processes in rat thymocytes. Proc Natl Acad Sci USA 68:1269–1273

Marver D (1980) Aldosterone receptors in rabbit renal cortex and rat medulla. Endo-crinology 106:611–618

McEwen BS (1980) Steroid hormones and the brain: cellular mechanisms underlying behavioral and neural plasticity. Psychoneuroendocrinology 5:1–11

McEwen BS, Luine VN (1978) Specificity, mechanism and functional significance of steroid-receptor interactions in the brain and pituitary. Coll Intern CNRS 280: 239–267

McEwen BS, Wallach G (1973) Corticosterone binding to hippocampus: nuclear and cytosol binding in vivo. Brain Res 57:373–386

McEwen BS, Weiss JM, Schwartz LS (1969) Uptake of corticosterone by rat brain and its concentration by certain limbic structures. Brain Res 16:227–241

McEwen BS, Zigmond RE, Azmitia EC Jr, Weiss JM (1970) Steroid hormone interaction with specific brain regions. In: Bowman RE, Datta SP (eds) Biochemistry of brain and behavior. Plenum Press, New York pp 123–167

McEwen BS, Wallach G, Magnus C (1974) Corticosterone binding to hippocampus: immediate and delayed influence of the absence of adrenal secretion. Brain Res 70:321–334

McEwen BS, De Kloet ER, Wallach G (1976) Interactions in vivo and in vitro of corticoids and progesterone with cell nuclei and soluble macromolecules from rat brain regions and pituitary. Brain Res 105:129–136

McEwen BS, Krey LC, Luine VN (1978) Steroid hormone action in the neuroendocrine system: When is the genome involved. In: Reichlin S, Baldessarini RJ, Martin JB (eds) The hypothalamus. Raven Press, New York, pp 255–266

McEwen BS, Stephenson BS, Krey LC (1980) Radioimmunoassay of brain tissue and cell nuclear corticosterone. J Neurol Sci Meth 3:57–65

McGaugh JL, Zornetzer SF, Gold PE, Landfield PW (1972) Modification of memory systems: some neurobiological aspects. Q Rev Biophys 5:163–186

McGinnis JF, De Vellis J (1978) Glucocorticoid regulation in rat brain cell cultures. J Biol Chem 253:8483–8492

McIntyre DC (1976) Adrenalectomy: protection from kindled convulsion induced amnesia in rats. Physiol Behav 17:789–795

McIntyre DC, Wann PD (1978) Adrenalectomy: protection from kindled convulsion induced dissociation in rats. Physiol Behav 20:469–474

McWilliams R, Lynch G (1980) Terminal proliferation in the partially deafferentiated dentate gyrus: time courses for the appearance and removal of degeneration and the replacement of lost terminals. J Comp Neurol 187:191–198

Meyer JS, Leveille PJ, McEwen BS, De Vellis J (1978) Glucocorticoids and glial cells: glycerolphosphate dehydrogenase induction and cytosol binding in normal and degenerated rat optic nerve. Annual Meeting of the Endocrine Society (Abstr 406)

Micco DJ Jr, McEwen BS (1980) Glucocorticoids, the hippocampus, and behavior: interactive relation between task activation and steroid hormone binding specificity. J Comp Physiol Psychol 94:624–633

Micco DJ Jr, McEwen BS, Shein W (1979) Modulation of behavioural inhibition in appetitive extinction following manipulation of adrenal steroids in rats: implications for involvement of the hippocampus. J Comp Physiol Psychol 93:323–329

Milgrom E, Luu Thi M, Baulieu EE (1973) Control mechanisms of steroid hormone receptors in the reproductive tract. Acta Endocrinol (Suppl) 180:380–403

Millard SA, Costa E, Gal EM (1972) On the control of brain serotonin and turnover rate by end product inhibition. Brain Res 40:545–551

Miller AL, Chaptal C, McEwen BS, Peck EJ Jr (1978) Modulation of high affinity GABA uptake into hippocampal synaptosomes by glucocorticoids. Psychoneuroendocrinology 3:155–164

Moguilevsky M, Raynaud JP (1980) Evidence for a specific mineralocorticoid receptor in rat pituitary and brain. J Steroid Biochem 12:309–314

Moyer KE (1958) Effect of adrenalectomy on anxiety motivated behavior. J Genet Psychol 92:11–16

Moyer KE (1966) Effect of ACTH on open-field behavior, avoidance, startle, and food and water consumption. J Genet Psychol 108:297–302

Moyer KE, Moshein P (1963) Effect of adrenalectomy on the attenuation of a conditioned avoidance response by ECS in the rat. J Comp Physiol Psychol 56:163–166

Nakajima S (1975) Amnesic effect of cycloheximide in the mouse mediated by adrenocortical hormones. J Comp Physiol Psychol 88:378–385

Nakajima S (1978) Attenuation of amnesia by hydrocortisone in the mouse. Physiol Behav 20:607–611

Nakanishi S, Kita T, Taii S, Imura H, Numa S (1977) Glucocorticoid effect on the level of corticotropin messenger RNA activity in rat pituitary. Proc Natl Acad Sci USA 74:3283–3286

Neckers L, Sze PY (1975) Regulation of 5-hydroxytryptamine metabolism in mouse brain by adrenal glucocorticoids. Brain Res 93:123–132

Nyakas C, De Kloet ER, Bohus B (1979) Hippocampal function and putative corticosterone receptors: Effect of septal lesions. Neuroendocrinology 29:301–312

Nyakas Cs, De Kloet ER, Veldhuis HD, Bohus B (1981) Corticosterone binding capacity increases in contralateral hippocampus after partial unilateral hippocampectomy. Neurosci Lett 21:339–343

Olpe H-R, McEwen BS (1976) Glucocorticoid binding to receptor-like proteins in rat brain and pituitary: ontogenetic and experimentally induced changes. Brain Res 105:121–128

Papez JW (1937) A proposed mechanism of emotion. Arch Neurol Psychiatr 38:725–744

Pappas BA, Gray P (1971) Cue value of dexamethasone for fearmotivated behavior. Physiol Behav 6:127–130

Pardridge WM, Mietus LJ (1979) Transport of steroid hormones through the rat blood brain barrier. J Clin Invest 64:145–154

Paul C, Havlena J (1962) Maze learning and open field behavior of adrenalectomized rats. J Psychosom Res 6:153–156

Pietras RJ, Szego CM (1977) Specific binding sites for oestrogen at the outer surface of isolated endometrial cells. Nature 265:69–72

Przewlocki R, Höllt V, Voigt KH, Herz A (1979) Modulation of in vitro release of β-endorphin from the separate lobes of the rat pituitary. Life Sci 24:1601–1608

Rees HD, Gray HF (1982) Glucocorticoids and mineralocorticoids: Actions on brain and behavior. In: Nemeroff CB, Dunn AJ (eds) Behavioral endocrinolgy. Spectrum Publ, New York, in press

Rees HD, Stumpf WE, Sar M (1975) Autoradiographic studies with ³H-dexamethasone in the rat brain and pituitary. In: Stumpf WE, Grant LD (eds) Anatomical neuroendocrinology. Karger, Basel, pp 262–269

Rees HD, Stumpf WE, Sar M, Petrusz P (1977) Autoradiographic studies of ³H-Dex uptake by immunocytochemically characterized cells of the rat pituitary. Cell Tissue Res 182:347–356

Rhees RW, Grosser BI, Stevens W (1975) Effect of steroid competition and time on the uptake of [³H]-corticosterone in the rat brain; an autoradiographic study. Brain Res 83:293–300

Rigter H, Crabbe JC (1979) Modulation of memory by pituitary hormones and related peptides. Vitam Horm 37:153–241

Robustelli F, Geller A, Jarvik ME (1972) Systematic analysis of the detention phenomenon in mice. J Comp Physiol Psychol 81:472–482

Rossier J, French E, Gros C, Minick S, Guillemin R, Bloom FE (1979) Adrenalectomy, dexamethasone or stress alters peptide levels in rat anterior pituitary but not intermediate lobe or brain. Life Sci 25:2105–2112

Roth GS (1976) Reduced glucocorticoid binding site concentration in cortical neuronal perikarya from senescent rats. Brain Res 107:345–354

Rousseau GG, Baxter JD, Tomkins GM (1972) Glucocorticoid receptors: relations between steroid binding and biologic effects. J Mol Biol 67:99–107

Rousseau GG, Baxter JD, Higgins SJ, Tomkins GM (1978) Steroid induced nuclear binding of glucocorticoid receptors in intact hepatoma cells. J Mol Biol 79:539–554

Sakly M, Koch B (1981) Ontogenesis of glucocorticoid receptors in anterior pituitary gland: Transient dissociation among cytoplasmic density, nuclear uptake and regulation of corticotropic activity. Endocrinology 108:591–596

Samuels HH, Tomkins GM (1970) Relation of steroid structure to enzyme induction in hepatoma tissue culture cells. J Mol Biol 52:57—74

Schapiro S (1968) Some physiological, biochemical and behavioral consequences of neonatal hormone administration: cortisol and thyroxine. Gen Comp Endocrinol 10:214—218

Selye H (1950) Stress. The physiology and pathology of exposure to stress. Acta Medica Publication, Montreal

Sherman MR, Pickering LA, Rollwagen FM, Miller LK (1978) Mero-receptors: proteolytic fragments of receptors containing the steroid-binding site. Fed Proc 37:167—173

Silva MTA (1974) Effects of adrenal demedullation and adrenalectomy on an active avoidance response of rats. Physiol Psychol 2:171—174

Simantov R (1979) Glucocorticoid inhibits endorphin synthesis by pituitary cells. Nature 280:684—685

Smelik PG, Papaikonomou E (1973) Steroid-feedback mechanisms in pituitary-adrenal function. Progr Brain Res 34:99—110

Squire LR, John SSt, Davis HP (1976) Inhibitors of protein synthesis and memory: dissociation of amnesic effects and effects on adrenal steroid genesis. Brain Res 112:200—206

Stevens W, Reed DJ, Erickson S, Grosser BI (1973) The binding of corticosterone to brain proteins, diurnal variation. Endocrinology 93:1152—1156

Stewart J, Krebs WH, Kaczender E (1967) State-dependent learning produced with steroids. Nature 216:1223—1224

Strum JM, Feldman D, Taggart B, Marver D, Edelman IS (1975) Autoradiographic localization of corticosterone receptors (Type III) to the collecting tubule of the rat kidney. Endocrinology 97:505—516

Stumpf WE, Sar M (1975) Anatomical distribution of corticosterone concentrating neurons in rat brain. In: Stumpf WE, Grant LD (eds) Anatomical neuroendocrinology. Karger, Basel, pp 254—261

Svec F, Harrison RW (1979) The intracellular distribution of natural and synthetic glucocorticoids in the AtT-20 cell. Endocrinology 104:1563—1568

Sze PY, Neckers L, Towle AC (1976) Glucocorticoids as a regulatory factor for brain tryptophan hydroxylase. J Neurochem 26:169—173

Tang F, Philips JG (1978) Pituitary-adrenal response in male rats subjected to continuous ether stress and the effect of dexamethasone treatment. J Endocrinol 75:325—326

Turner BB (1978) Ontogeny of glucocorticoid binding in rodent brain. Am Zool 18:461—475

Turner BB, McEwen BS (1980) Hippocampal cytosol binding capacity of corticosterone: no depletion with nuclear loading. Brain Res 189:169—182

Vale W, Rivier C, Yang L, Minick S, Guillemin R (1978) Effects of purified hypothalamic corticotropin-releasing factor and other substances on the secretion of adrenocorticotropin and β-endorphin-like immunoactivities in vitro. Endocrinology 103:1910—1915

Valeri P, Angelucci L, Palmery M (1978) Specific ^{3}H-corticosterone uptake in the hippocampus and septum varies with social settings in mice. Neurosci Lett 9:249—254

Valtin H, Schroeder HA (1964) Familial hypothalamic diabetes insipidus in rats (Brattleboro strain). Am J Physiol 206:425—530

Van Delft AML (1970) Conditioned avoidance behavior and the pituitary-adrenal system in the rat. Ph D Thesis, University of Utrecht

Van Dijk AMA, Van Wimersma Greidanus TjB, Burbach JPH, De Kloet ER, De Wied D (1981) Brain adrenocorticotrophin after adrenalectomy and sham-operation of rats. J Endocrinolog 88:243—253

Van Wimersma Greidanus TjB (1970) Effects of steroids on extinction of an avoidance response in rats. A structure-activity relationship study. Progr Brain Res 32:185–191

Van Wimersma Greidanus TjB, De Wied D (1969) Effects of intracerebral implantation of corticosteroids on extinction of an avoidance response in rats. Physiol Behav 4: 365–370

Van Wimersma Greidanus TjB, De Wied D (1971) Effects of systemic and intracerebral administration of two opposite acting ACTH-related peptides on extinction of conditioned avoidance behavior. Neuroendocrinology 7:291–301

Van Wimersma Greidanus TjB, Wijnen H, Deurloo J, De Wied D (1973) Analysis of the effect of progesterone on avoidance behavior. Horm Behav 4:19–30

Van Wimersma Greidanus TjB, Bohus B, De Wied D (1974) Differential localization of the influence of lysine-vasopressin and of ACTH 4–10 on avoidance behavior: a study in rats bearing lesions in the parafascicular nuclei. Neuroendocrinology 14: 280–288

Veldhuis D, De Kloet ER (1981) Capacity of corticosterone receptor system in rat brain: Control by neuropeptides and hormones. In: Stark E, Makara GB, Acs Zs, Endröczi E (eds) Proceedings of the XXVIII Intern Congress of Physiological Sciences, vol 13. Akademiai Kiado, Budapest, pp 61–65

Veldhuis HD, De Kloet ER (1982a) Significance of ACTH 4–10 in the control of hippocampal corticosterone receptor capacity of hypophysectomized rats. Neuroendocrinology

Veldhuis HD, De Kloet ER (1982b) Vasopressin-related peptides increase the hippocampal corticosterone receptor capacity of diabetes insipidus (Brattleboro) rats. Endocrinology 110:153–157

Vermes I, Smelik PG, Mulder AH (1976) Effects of hypophysectomy, adrenalectomy and corticosterone treatment on uptake and release of putative central neurotransmitters by rat hypothalamic tissue in vitro. Life Sci 19:1719–1726

Versteeg DHG, De Kloet ER, Van Wimersma Greidanus TjB, De Wied D (1979) Vasopressin modulates the activity of catecholamine containing neurons in specific brain regions. Neurosci Lett 11:69–73

Von Zerssen D (1976) Mood and behavioural changes under corticosteroid therapy. In: Itil TM, Laudahn G, Herrmann WM (eds) Psychotropic action of hormones. Spectrum Publ, New York, pp 195–222

Warembourg M (1975) Radioautographic study of the rat brain after injection of $[1,2-^3H]$ corticosterone. Brain Res 89:61–70

Watson SJ, Akil H (1981) On the multiplicity of active substances in single neurons: β-endorphin and α-MSH as a model system. In: De Wied D, van Keep PA (eds) Hormones and the brain. MTP Press, Lancaster, pp 73–86

Watson S, Akil H, Richard CW, Barchs JD (1978) Evidence for two separate opiate peptide neuronal systems. Science 275:226–228

Weiss JM, McEwen BS, Silva MT, Kalkut M (1970) Pituitary-adrenal alterations and fear responding. Am J Physiol 218:864–868

Wertheim GA, Conner RL, Levine S (1967) Adrenocortical influences on free-operant avoidance behavior. J Exp Anal Behav 10:555–563

Westphal U (1971) Steroid protein interaction. Springer, Berlin Heidelberg New York

Weijnen JAWM, Slangen JL (1970) Effects of ACTH-analogues on extinction of conditioned behavior. Progr Brain Res 32:221–235

Woodbury DM (1972) Biochemical effects of adrenocortical steroids on the central nervous system. In: Lajtha A (ed) Handbook of Neurochemistry, vol VII. Plenum Press, New York, pp 225–287

Wrange O (1979) A comparison of the glucocorticoid receptor in cytosol from rat liver and hippocampus. Biochim Biophys Acta 582:346–357

Subject Index

Current Topics in Neuroendocrinology

The central nervous system and the endocrine system comprise quickly expanding fields at the forefront of medical and as biological research. Neuroendocrinology is pursued so intensely that it has become impossible to keep an up-to-date view of all current research articles. The purpose of *Current Topics in Neuroendocrinology* is to treat topics which are also reliable where enough work has been done to treat them from several aspects.

Volume 1

Sleep

Clinical and Experimental Aspects

1982. 47 figures. 136 pages
ISBN 3-540-11125-5

Contents: Neurobiology of REM Sleep. A Possible Role for Dopamine. – Endocrine and Peptide Functions in the Sleep-Waking Cycle. – Sleep Regulation: Circadian Rhythm and Homeostasis. – Haemodynamic Changes during Sleep in Man. – Subject Index.

Volume 1 covers mechanisms of **Sleeps** from a neurochemical, endocrine and clinical point of view. In four chapters, in-depth discussions treat neurotransmitters, pharmacologic interference, endocrine and peptidergic functions, a characterization of sleep with regard to hemodynamics, single unit recording and EEG activity as well as biologic rhythms and sleep characteristics in patients.

Springer-Verlag
Berlin
Heidelberg
New York

P.Böck

The Paraganglia

1982. 61 figures, approx. 43 tables.
Approx. 368 pages
(Handbuch der mikroskopischen Anatomie des
Menschen, Volume 6, Part 8)
ISBN 3-540-10978-1

Gonadal Steroids and Brain Function

IUPS-Satellite Symposium, Berlin, July 10–11, 1980
Editors: W.Wuttke, R.Horowski
1981. 136 figures, 10 tables. XIII, 373 pages
ISBN 3-540-10606-5

Glucocorticoid Hormone Action

Editors: J.D.Baxter, G.G.Rousseau
1979. 176 figures, 58 tables. XIX, 638 pages
(Monographs on Endocrinology, Volume 12)
ISBN 3-540-08973-X

A.Labhart

Clinical Endocrinology

Theory and Practice
With a Foreword by G.W.Thorn
In collaboration with numerous experts.
Translators: A.Trachsler, J.Dodsworth-Phillips
1974. 400 figures. XXXII, 1092 pages
ISBN 3-540-06307-2

A.M.Neville, M.J.O'Hare

The Human Adrenal Cortex

Pathology and Biology – An Integrated Approach
1982. 173 figures. XIV, 354 pages
ISBN 3-540-11085-2

The Renin Angiotensin System

A Model for the Synthesis of Peptides in the Brain
Editors: D.Ganten, M.P.Printz, M.I.Phillips,
B.A.Schölkens
1982. 87 figures, 55 tables.
(Experimental Brain Research, Supplementum 4)
ISBN 3-540-11344-4

Springer-Verlag
Berlin
Heidelberg
NewYork

GPSR Compliance
The European Union's (EU) General Product Safety Regulation (GPSR) is a set
of rules that requires consumer products to be safe and our obligations to
ensure this.

If you have any concerns about our products, you can contact us on

ProductSafety@springernature.com

In case Publisher is established outside the EU, the EU authorized
representative is:

Springer Nature Customer Service Center GmbH
Europaplatz 3
69115 Heidelberg, Germany